Schooling for Sustainable Development

Volume 7

This book series addresses issues associated with sustainability with a strong focus on the need for educational policy and action. Current attention and initiatives assume that Education for Sustainable Development (ESD) can be introduced successfully and gradually into schools worldwide. This series explores the issues that arise from the substantial and sustainable changes to be implemented in schools and education systems.

The series aims to counter the prevailing Western character of current research and enable cross-cultural comparisons of educational policy, practice, and project development. As a whole, it provides authoritative and comprehensive global coverage, with each volume providing regional/continental coverage. The volumes present data and insights that contribute to research, policy and practice in ESD-related curriculum development, school organization and school-community partnerships. They are based on ESD-related project experiences, empirical studies that focus on ESD implementation and teachers' perceptions as well as childhood studies that examine children's geographies, cultural characteristics and behaviours.

More information about this series at http://www.springer.com/series/8635

Rajeswari Namagiri Gorana
Preeti Rawat Kanaujia
Editors

Reorienting Educational Efforts for Sustainable Development

Experiences from South Asia

Springer

Editors
Rajeswari Namagiri Gorana
Children's Media Unit
Centre for Environment Education (CEE)
Ahmedabad, India

Preeti Rawat Kanaujia
Northern Regional Cell
Centre for Environment Education (CEE)
Lucknow, India

Schooling for Sustainable Development
ISBN 978-94-024-1396-0 ISBN 978-94-017-7622-6 (eBook)
DOI 10.1007/978-94-017-7622-6

Printed on acid-free paper

This Springer imprint is published by Springer Nature
The registered company is Springer Science+Business Media B.V. Dordrecht

To
our daughters

Gauri and Prakriti

for sharing their time
for being ever supportive

Foreword: Education as a Key Driver of Change

As the UN Decade of Education for Sustainable Development (DESD) began in 2005 experts, NGOs, media, the private sector, practitioners, students and citizens from all over the world gathered in Ahmedabad to consider the challenge ahead and how they could play a positive role as part of the conference *Education for a Sustainable Future*. The conference led to the Ahmedabad Declaration of 2005 which said, "We firmly believe that a key to sustainable development is the empowerment of all people, according to the principles of equity and social justice, and that a key to such empowerment is action-oriented education."

In 2014, as the DESD drew to a close, education, the "E" in ESD seemed to be generally better understood. The year 2015 saw a better definition of the "SD" in ESD. In September 2015, for the first time, 193 nations of the world came together at the UN and concurred on a working definition of Sustainable Development. At this meeting, 17 goals and 169 targets were recognized as essential for sustainable development. 2015 was also the year of the Paris Agreement when 160 countries decided on their Nationally Determined Commitments (NDCs) for combating climate change at CoP 21 of the United Nations Framework Convention on Climate Change.

We now need to focus on achieving these goals. This is when strategies are being formed for climate change and more widely for sustainable development. It was therefore important to bring the lessons of the DESD and others to share these in a way that they would be useful to planners and other decision makers who have the potential to integrate education into their strategies. The international conference *Education as a Driver of Sustainable Development Goals* in Ahmedabad, India, organized by CEE in partnership with UNESCO, UNEP, Ministry of Environment, Forest and Climate Change and NITI Aayog (National Institution for Transforming India), Government of India, was a timely discussion coming as it did after the goal setting and before the planning of strategies.

It has taken more than four decades for the world to acknowledge the power and potential of 'schooling for sustainability', which is the theme of this book series. This series address issues associated with sustainability with a strong focus on the need for educational policy and action. Environment largely focused on pollution and wildlife when the world community met in Stockholm in 1972 for the United Nations Conference on the Human Environment. And, it was not really until 1992 in Rio de Janeiro at the United Nations Conference on Environment and Development that the urgency and importance of recognizing – environment and development – as twin goals got established.

Countries of South Asia largely sharing a syncretic culture, colonial past and common developmental challenges need to question 'given' wisdom, paradigms and worldviews. The challenge before developing countries is essentially different from that before developed nations. Accordingly, the education strategies and goals are also different.

To illustrate, in 2008, United Nations Environment Programme which coordinates the celebration of the World Environment Day each year on June 5th selected the slogan *CO_2: Kick the Habit* in response to the growing climate change crises. In India where electricity connections may not even exist in remote areas, this seemed a rather strange slogan. Were the people who did not have access to electricity supposed to first form the '*CO_2*' habit and then 'kick' it? We felt, in India, that we needed to change the 'CO_2: Kick the Habit' campaign to what would be more appropriate for a developing nation. Thus emerged the *CO_2: Pick Right* campaign, so the focus was on making the right choices rather than imitating outdated solutions. It identified the need to 'leapfrog' developing countries straight into sustainable development without first having to make the mistakes of the early developers.

But what is, perhaps, not as widely recognized is that it is not only the paradigm of development that needs to be changed to achieve sustainability, but that the paradigm of education also has to change in certain fundamental ways. Education needs to strongly focus on critical thinking, the ability to take positive action and to see a problem in its holistic perspective.

In 2005, Srija, a 10-year old student participating in one of the CEE's school programmes, offered her handprint as a commitment to action for sustainability. Subsequently, the handprint became the symbol of sustainability. The handprint is a measure today which stands for caring and working together towards a sustainable future. It is important to give children a positive message about their ability to effect positive environmental change.

Diverse sectors and environmental issues need ESD but the approaches need to be locally effective and appropriate. There are numerous examples across South Asia which demonstrate that education can be a key driver for change that is environmentally sound, economically viable and socially beneficial.

This book focuses on initiatives of ESD and the challenges, opportunities, issues and strategies in Bangladesh, Bhutan, India, Maldives, Nepal, Pakistan and Sri Lanka. The authors have examined, where necessary, the administrative, economic, social, cultural and ecological realities at various levels of policy, planning, implementation and evaluation. The discussions in the book extend beyond formal education systems like schools, higher education, pre-service and in-service teacher preparation to community education, communication and awareness initiatives conceptualized for realizing environmental, social and economic goals of sustainability.

Initiatives presented in the chapters demonstrate how each country, in its own way, is contributing to operationalize ESD to move towards sustainable development. To quote the Ahmedabad Declaration of 2005, once again, "We pledge to share our diverse experiences and collective knowledge to refine the vision of sustainability while continually expanding its practice. Through our actions we will add substance and vigor to the UNDESD processes."

I am sure this book would be an important reference for ESD professionals, policy makers, researchers, practitioners and people working in the region and at the global level while enriching the prevailing paradigms of ESD thinking and providing insights to enable cross-cultural comparisons of ESD practice.

Founder Director, CEE, India Kartikeya V. Sarabhai
January 2016

Series Editors' Introduction

Education for sustainable development (ESD) has rapidly become part of educational discourses worldwide. Within its global attractiveness lie both its strength and its weakness. Its strength lies in its capacity to alert educationists, broadly defined, to a shared concern for the future of both the planet and local communities. Its weakness lies in its lack of shared meaning and, stemming from this, the enormous difficulties encountered in trying to bring ESD into the mainstream activities of educational institutions. The period 2005–2014 was designated the United Nations Decade of Education for Sustainable Development during which the United Nations sought to bring to the fore the need for politicians, policy makers and practitioners to seek ways by which ESD can become part of the fabric of formal and informal education. At the heart of the numerous initiatives that have been stimulated by this designation is the assumption that ESD should be introduced and can be introduced successfully into schools worldwide. It is assumed that children, older students and adults can be educated formally and informally to act now in the interests of a sustainable future and to act internationally.

What is evident is that different nations have adopted different approaches to ESD, sometimes interchanging the term with environmental education, another term subject to a wide range of interpretations. These differences are evident in educational practice in regions, districts and individual schools as well as in academic studies and commentaries. Obviously, this is not to say that there is some common ground in policies and practice, it is simply to keep to the forefront the recognition that, even when nations make pronouncements about aspects of ESD, these should not be treated as authoritative statements about what is happening at the school and classroom levels. Broad statements have a value in highlighting issues and trends but they need to be treated with caution. The same caution needs to be applied to pronouncements emanating from academic sources. Academics have their own agendas and care must be taken when reading what appear to be authoritative statements about developments in ESD occurring within their own communities and nations.

Our series addresses the array of issues arising from attempts made to convert assumptions about, and definitions of, ESD into substantial and sustainable changes principally in schools. Underpinning the series is a concern for identifying those cultural forces that impact on national, regional and local adaptations to approaches to ESD that have international currency. In this, the editors of the books in the series, each based on experience in a single continent or extensive region, seek to counter the strong Western (Australian, North American, European) character of much research and writing in the broad field of ESD. Research and scholarly studies are commonly underpinned by values and assumptions derived from Western culture, broadly defined. The design of the series as a set of broadly continent-scale books seeks to bring together experts from various countries in each continent. The books bring out contrasting experiences and insights with a range of explanations of policies and practice.

Within the broad cultural contexts of the continents and regions included in the series authors provide evidence of policies, formal curriculum developments and innovations and informal school-related activities. Some authors have paid close attention to policy making at various levels, others have addressed whole school organizational issues and others have provided detailed case studies of localities and individual schools.

Children and young people live in distinct worlds of their own. They have very distinctive cognitive and affective characteristics that vary from one culture to another, at whatever scale that culture is defined. They are also often targets for environmental campaigns that wish to promote particular behavioural changes. ESD is often construed as an attempt to change habits, to encourage children and young people to "think globally and act locally". This series demonstrates how this and other slogans are translated in education systems and schools worldwide.

For this volume *Reorienting Educational Efforts for Sustainable Development: Experiences from South Asia*, the co-editors Rajeswari Namagiri Gorana and Preeti Rawat Kanaujia have brought together an array of chapters highlighting the recent developments and issues related to ESD in seven South Asian nations.

December 2015

John Chi-Kin Lee
Michael Williams
Philip Stimpson

Acknowledgements

It was enriching for us to work on this book – difficult at times, but always a learning experience. The lessons learnt during this journey will be with us forever, for which we have to acknowledge many.

We begin with our organization, the Centre for Environment Education (CEE), for having given us a practical and holistic understanding of EE and ESD. Each day at CEE and all of its initiatives contributed to this understanding. They helped us in our conversations with the authors.

We express our sincere gratitude to Kartikeya V. Sarabhai, the founding Director of CEE, for giving us the space and support to work on the book. He is an inspiration and his vision of ESD has guided us.

CEE's stature in the region made it possible to contact the authors and seek their contribution. Thanks to the book, we have got to know these fine people. Many thanks are due to each one of them for showing phenomenal patience. They reminded us of the book now and again and hoped the best for it, in spite of delays.

We express our heartfelt gratitude to Mamata Pandya, our guide and mentor, who believed in our abilities. We have benefited in many ways from her support and encouragement. Importantly, we would like to thank Kiran Chhokar and M J Ravindranath who encouraged us to work on the book.

We thank our colleagues M J, Ravindranath, Sanskriti Menon, Ramesh Savalia, Sarita Thakore and Satish Awate who reviewed the contributions and gave valuable inputs.

For preparing the manuscript, we must thank our colleagues Namita Khare and JK Nair and our interns Priyanka Pandey and Ranjhani Iyer.

Heartfelt thanks to our Series Editors, Professor John Chi-Kin Lee, Professor Michael Williams and Dr. Philip Stimpson, for choosing CEE to bring out the South Asia book. As editors, we received their unwavering support which has been empowering for us.

Lastly, it is a pleasure for us to put on record that our awe and admiration for the people of South Asia increased manifoldly as we worked on the book. The region demonstrates very well that there are more interpretations than one to EE and ESD, which needs to be acknowledged.

Contents

Contributors

Badri Dev Pande teaches education for sustainable development (ESD) at the National College for Higher Education based in Kathmandu. He received his PhD in education from Southern Illinois University, USA. In addition to carrying out educational research, he taught in two different universities in Nepal. He coordinated the Environmental Education and Communication Programme of IUCN Country Office for over twelve years. He has been actively involved as a writer and contributor to workshops/symposiums on Education for Sustainable Development in Asia. He has several publications focusing on environmental education and ESD.

Tahir Ahmad Dhindsa is founder managing director of Sustainable Development Television (sdpi.tv), the pioneering Web-TV in the region and the Editor of *Sustainable Development Policy Economic Bulletin*, published by Sustainable Development Policy Institute, Islamabad. He has started his career as a journalist with the launch of the daily *The News Islamabad* in the early nineties and subsequently worked for the daily *The Muslim*, *Dawn*, the weekly *Tribune* and the daily *The Business Recorder*. He worked for Pakistan Television and Geo News as news producer. He has also worked for CNBS and News Asia channels as a foreign correspondent. In the development sector, Tahir A. Dhindsa has worked for the World Bank and UNDP as a communications specialist; for many years, he had been part of the team responsible for the 'Pakistan Country Partnership Report' of the World Bank. He has also worked as head of the Investors' education department in the Securities and Exchange Commission of Pakistan. In addition, he has taught in many government and private universities in Islamabad. He holds a postgraduate degree in English literature from the University of the Punjab, Lahore (Pakistan).

Madhavi Joshi is Programme Director at the Centre for Environment Education, Ahmedabad. She worked in the Secretariat for the Decade of Education Sustainable Development (DESD) in India helping the National Nodal Agency, Ministry of Human Development (MHRD), developing National DESD Activities in India along with documenting DESD progress across the world through a dedicated website. She conducted the study and co-authored the chapter on South Asia Situational Analysis of ESD as a part of the Asia Pacific Situational Analysis of ESD anchored by the UNESCO Asia Pacific Bureau. She has been involved in the DESD Asia Pacific strategy formulation and the key capacity building workshops held for the South Asia region. She initiated the Youth Initiatives at the Centre which have grown into a South Asia level programme. She currently coordinates the South Asia Youth Environment Network (SAYEN), a UNEP TUNZA Children and Youth Programme. As a part of the youth programmes, her interest is in creating learning opportunities for youth through action programmes such as the Handprint Campus. She is engaged in urban issues especially in the area of facilitating collaborative and participatory dialogue among multi-stakeholder groups. Currently, she co-teaches a postgraduate course on urban mobility and people at the CEPT University, Ahmedabad. She also teaches Environmental Studies subject to undergraduates at the Ahmedabad University.

Preeti Rawat Kanaujia is a Senior Programme Coordinator at the northern regional office of the Centre for Environment Education (CEE) in India. She holds a Master's degree in environmental sciences. She has been involved in various formal and non-formal school education initiatives in the field of environmental education and education for sustainable development at the CEE. She has published widely in the fields of teachers' educational materials and in school improvement and educational evaluation. She has coordinated national level environmental education and education for sustainable development initiatives. She facilitates and participates in international ESD network Regional Centres of Expertise anchored by United Nations University – Institute of Advanced Studies.

Rajeswari Namagiri Gorana is a Programme Coordinator at the Children's Media Unit of the Centre for Environment Education (CEE), India. She holds a Master's degree in environmental sciences and a Bachelor's degree in education from Andhra University. At CEE, she is involved in developing education and communication materials on environmental education and education for sustainable development. She has developed a variety of teaching and learning materials. Her recent publication *Shabab Al Estidama: A Resource for Youth on Sustainability*, co-authored with Mamata Pandya and Gayatri Raghwa, was for youth.

She has designed an e-Course titled *Certificate course in Educating for Sustainability* for in-service teachers at the intermediate level for Environment Agency – Abu Dhabi.

Her interests include developing environmental games and online learning. She is a member of the ESD ExpertNet, a trans-national (India, Germany, Mexico and South Africa) network of experts which enhances the implementation of ESD in the network countries.

Mamata Pandya has served as a Senior Programme Director in the Instructional Design Unit at the Centre for Environment Education (CEE), India. She has a Masters degree in Political Science from the University of Delhi, and in South Asian Studies from the University of Sussex, UK. She has been involved in the development of environmental education materials and programmes for teachers, students and the general public covering a wide range of media including print, exhibitions, videos and new media. She has been a member of the development team for multi-media educational packages under the PLANET programme of the Asia Pacific Cultural Centre for UNESCO, Japan, as well as a children's education package produced by the United Nations Environment Programme, Nairobi. She has been a part of India's National Council of Educational Research and Training's (NCERT) national text-book development committee which prepared the Environmental Studies (EVS) text-books for grades 3–5. Author, editor and co-editor of over a hundred publications for teachers and children, Mamata is now an independent consultant and editor.

Shailaja Ravindranath is Regional Director of the southern regional office of the Centre for Environment Education (CEE) in India. She coordinates programmes in the area of Environmental Education and Education for Sustainable Development in the Southern States of India. She has a PhD in plant biochemistry from the M S University of Baroda. After completing her PhD, she spent a year in the Central Plantation Crops Research Institute and a year in Karnatak University, Dharwad, working on physiological aspects of plants using in vitro techniques. However, her interest shifted from laboratory studies to development related field studies when she joined the Indian Institute of Science, Bangalore. Scholars and great teachers like Prof. Amulya Kumar N. Reddy and Prof. K.S. Jagdish supported and encouraged her interests there. While working with the rural community, especially women, she realized the importance of education to achieve the goals of sustainable development. She then moved to CEE to pursue her interests. She now is working towards integrating the principles and the processes of Education for Sustainable Development (ESD) in development interventions.

Md. Salequzzaman is Professor and Founding Head of the Environmental Science Programme of Khulna University, Bangladesh. His PhD was awarded by Murdoch University (Western Australia) and focused on sustainable technology and policy. His research interests include appropriateness of disaster management, waste management, integrated coastal zone management, sustainable aquaculture, waste minimization and resource recovery, treatment of wastewater and landfill leachate and renewable energy (tidal, biogas and solar). He has several publications on the environment and sustainable development.

Upali M. Sedere is an Education Sector Specialist and former Director General of the National Institute of Education, Sri Lanka. His doctorate in Education was awarded by the University of Iowa, USA, where he had a Fulbright scholarship. His areas of specialization are social sector policy formulation, large-scale programme and project management and implementation, capacity building; planning, monitoring and evaluation; programme quality improvement, project appraisal and supervision. He worked as Head of Department and Senior Lecturer for eighteen years in the Faculty of Education in Sri Lanka. He has been listed in the American Biographical Institute's International 'Who's Who' for 2003. He has several publications and wide experience of working in various countries.

Mohamed Shareef is Lecturer at the Faculty of Education, Maldives National University, which is the largest tertiary education institution in the Maldives. He is currently teaching Environmental Education at the Department of Social Sciences of the Faculty. In addition, he has developed unit outlines and reading materials and has coordinated the University of Brunei Darussalam with a Bachelor of Arts Education (Honours) degree. He has special academic interests in teaching and research in Geography, Environmental Studies, Social Studies and History. He has made a significant contribution to the understanding of Environmental Education in the Maldives through research he undertook in his Masters degree at the Unitec Institute of Technology, Auckland, New Zealand. Previously Mr. Shareef taught Advanced Level Geography, at the Centre for Higher Secondary Education, which is also the largest higher secondary education institution in the Maldives.

Dasho Karma Ura is President of the Centre for Bhutan Studies (CBS) in Bhutan. He studied for an undergraduate degree at Oxford University and he was awarded a Masters degree in Philosophy of Economics from the University of Edinburgh. After working for the Ministry of Planning for twelve years, he served as the Director of CBS since its establishment and became its President in 2008. Under his leadership, CBS has been at the forefront in deepening national and global understanding of Bhutan's development philosophy of Gross National Happiness. He was a member of the Drafting Committee of Bhutan's first Constitution and holds numerous international positions. He has been awarded many honours including the Red Scarf and the ancient title of Dasho (knighthood) by His Majesty the Fourth King. He is also a renowned painter and has written several books.

Ramboojh Yadav is a Programme Specialist in Ecological and Earth Sciences at UNESCO's office, New Delhi. He is a focal point for the Man and Biosphere (MAB) Programme in South Asia. He is Secretary to the South and Central Asian Network of the Man and Biosphere (SACAM) programme of UNESCO. He has been active in the UN Decade of Education for Sustainable Development (DESD) and in coordinating World Heritage (Natural), Climate Change Knowledge (CLICK) Partnership programmes of UNESCO.

Acronyms

ACCU	Asia Pacific Cultural Centre for UNESCO
AOSIS	Alliance of Small Island States
BMZ	The Federal Ministry for Economic Cooperation and Development, Germany
BVIEER	Bharati Vidyapeeth Institute for Environmental Education and Research
CBOs	Community-Based organizations
CBS	Centre for Bhutan Studies
CC:iNet	Climate Change Information Network
CEE	Centre for Environment Education
CEEP	Coordinated Environmental Education Project
CMZ	Coastal Management Zone
COE	Centres of Excellence
CSR	Corporate Social Responsibility
DAC	Development Assistance Committee
DAM	Dhaka Ahsania Mission
DANIDA	Danish International Development Agency
DEO	District Education Office
DESD	Decade of Education Sustainable Development
DPEP	District Primary Education Programme
EA	Environmental Authority
ECCA	Environmental Camps for Conservation Awareness
ECR	Education for Conflict Resolution
EDC	Educational Development Centre
EE	Environmental Education
EESS	Environmental Education in the School System
EFA	Education for All
EG	Engagement Global
EOSE	Environmental Orientation to School Education
EPSD	Education for Peace and Sustainable Development
ESC	Education for Social Cohesion Programme

ESD	Education for Sustainable Development
ESD-J	Japan Council on the UN Decade of Education for Sustainable Development
ESF	Education for a Sustainable Future
EVS	Environmental Studies
FINNIDA	Finnish Ministry of Foreign Affairs, Department for International Development Cooperation
FYPs	Five Year Plans
GCE	General Certificate of Education
GCSE	General Certificate of Secondary Education
GDP	Gross Domestic Product
GNH	Gross National Happiness
GoSL	Government of Sri Lanka
GTZ	German Technical Cooperation
ICEE	International Conference on Environmental Education
ICT	Information and communication technology
IED	Institute for Educational Development
IEEP	International Environmental Education Programme
IGES	Institute for Global Environmental Strategies
IIS	International Implementation Scheme
INGOs	International non-governmental organizations
IUCN	International Union for Conservation of Nature
LTTE	Liberation Tigers of Tamil Eelam
MAB	Man and Biosphere
MDGs	Millennium Development Goals
MEEW	Ministry of Environment, Energy and Water
MHRD	Ministry of Human Resource Development
MNU	Maldives National University
MOE	Ministry of Education
MoEF	Ministry of Environment and Forests
MoU	Memorandum of Understanding
NAPA	National Adaptation Programme of Action
NASSD	Northern Areas Strategy for Sustainable Development
NCEPC	National Committee on Environmental Planning and Coordination
NCERT	National Council of Educational Research and Training
NCF	National Curriculum Framework
NCFSE	National Curriculum Framework for School Education
NCS	National Conservation Strategy
NCSIP	National Conservation Strategy Implementation Project
NEAC	National Environmental Awareness Campaign
NEAP	National Environment Action Plan
NEMAP	National Environmental Management Action Plan
NEPC	National Educational Planning Commission
NFE	Non-formal Education
NFEC	Non-formal Education Centre

NGC	National Green Corps
NGO	Non-governmental Organizations
NIE	National Institute of Education
NORAD	Norwegian Agency for Development Cooperation
NPC	National Planning Commission
NPE	National Policy of Education
NSDS	National Sustainable Development Strategy
NUEPA	National University of Educational Planning and Administration
OECD	Organization for Economic Cooperation and Development
OECD DAC	Organisation for Economic Co-operation and Development: Development Assistance Committee
PLANET	Package Learning Materials on Environment
PRSP	Poverty Reduction Strategy Paper
RCEs	Regional Centres of Expertise
SACEP	South Asia Co-operative Environment Programme
SASEANEE	South and Southeast Asia Network for Environmental Education
SCERT	State Council of Educational Research and Training
SD	Sustainable development
SDPI	Sustainable Development Policy Institute
SHG	Self Help Groups
SIDS	Small Island Developing States
ToT	Teachers and teacher trainers
TVET	Technical and Vocational Education and Training
UGC	University Grants Commission
UNCED	United Nations Conference on Environment and Development
UNDESD	United Nations Decade of Education for Sustainable Development
UNDP	United Nations Development Programme
UNEP	United Nations Environment Programme
UNESCO	United Nations Educational, Scientific and Cultural Organization
UNFCCC	United Nations Framework Convention on Climate Change
UNICEF	United Nations Children's Fund
UNLD	United Nations Literacy Decade
WSSD	World Summit for Sustainable Development
WWF	Worldwide Fund for Nature

An Introduction to South Asia

Countries of the South Asia are connected by geography, history, cultural and religious heritage and by affiliations. They show some specific forms of conformity, requiring at the same time nuancing the popular belief that cultures are pan regional. The region far from being a homogenous entity is a mosaic cultures, in which people from different races, ethnicity and religions co-exist. The region houses all the major religions of the world: Hinduism, Buddhism, Christianity and Islam. The socio-political geography of the region is complex, and civil wars and sub-regional conflicts exist.

Bangladesh, Bhutan, India, Maldives, Nepal, Pakistan and Sri Lanka are the seven countries that make up South Asia. The region is also referred to as the Indian subcontinent.

> Historians Sugata Bose and Ayesha Jalal hold the view that the Indian Subcontinent has come to be known as South Asia 'in more recent and neutral parlance' though their view is not acceptable to many. Indologist Ronald B. Inden argues that the usage of the term 'South Asia' is getting more widespread since it clearly distinguishes the region from East Asia. Some academics hold that the term 'South Asia' is in more common use in Europe and North America, rather than the terms 'Subcontinent' or the 'Indian Subcontinent'. According to political science professor Tatu Vanhanen, 'the seven countries of South Asia constitute geographically a compact region around the Indian Subcontinent'; while according to anthropologist John R. Lukacs, 'the Indian Subcontinent occupies the major landmass of South Asia' (http://en.wikipedia.org/wiki/Indian_subcontinent. Accessed on May 6, 2013).

South Asia, both geographically and geopolitically, is a region of sharp extremes – whether it is biodiversity, population or governance in comparison to the other regions of the world.

Colonial Pasts

To understand South Asia's political, economic, governance, judicial and educational systems, it is important to look back into the colonial histories of the countries in the region. Portugal, France, the Netherlands and Great Britain colonized the countries at different times. "Asia is always an important place for Europe. It's been interested for at least 500 years in commerce, in resources, in its mineral wealth, and also in its people". (http://blogs.utexas.edu/15minutehistory/2013/01/23/episode-9-the-end-of-colonialism-in-south-asia/. Accessed on May 8, 2013). Except Nepal, the other six countries were under colonial rule or control for varying durations.

India acquired independence in 1947 after more than two hundred years, although significant portions of the Indian Territory were under the control of the East India Company for about one hundred years before the rule of the British Crown was established in 1858. Before 1947, India was a colony of Britain, so technically Pakistan and Bangladesh were under British rule. The British exerted some control over Bhutan's affairs, but never formally colonized it. "For centuries, Bhutan was made up of feuding regions until it was unified under King Ugyen Wangchuck in 1907" (http://cypresshimalaya.com/history-of-bhutan/. Accessed on December 9, 2015). Bhutan shifted to a constitutional monarchy in 2008 when the first national election took place. Sri Lanka was colonized in 1500 by the Portuguese and later in 1600 the Dutch arrived. However, in 1796, Dutch rule gave way to the British before becoming independent in 1948. (https://en.wikipedia.org/wiki/History_of_British_Ceylon. Accessed on December 9, 2015) The British ruled Maldives from 1887 to 1965 and before that it was colonized by the Portuguese and the Dutch. In 1965, Maldives gained independence under an agreement signed with Britain. (http://www.globalsecurity.org/military/world/indian-ocean/mv-history.htm. Accessed on December 9, 2015).

Colonization affected the countries of the region profoundly as it bequeathed them with a large legacy of the colonizer's systems. Often, the systems were to suit the imperialist systems of the colonizers and not so much the need of the colonized people.

> When India became independent, we were in a hurry to provide economic development for India's poor. We borrowed concepts, borrowed institutional models and borrowed models of economic development. We are anxious to catch up with the West, with its technologies and lifestyles. After 50 years of looking to the West for ideas, we have begun to discover that borrowed ideas do not work: And as Gandhiji put it 'we need to learn to write on the Indian Slate'. (Kamala Chowdhary as quoted by Kartikeya Sarabhai in his presentation titled *Securing our Future in the New Century: Lessons from India* made at the World Conservation Union Regional Conservation Forum held in 2000 at New Delhi, India) (http://www.ceeindia.org/newintra/papers/Securing%20our%20Future.htm.Accessed on May 6, 2013).

Colonial rule created a class of people that believed in the industrialization, education and modernization of the colonizers and wanted to emulate them.

> While colonial intrusion and the reorganization of native society to serve the requirements of European capital had certainly occasioned untold destruction, the corollary was that

> capitalism itself – with its railroads, industrial infrastructure, and communication systems – had introduced a dynamic historical germ that would rouse Indian society from the timeless stagnation of the Asiatic mode of production and set it on its own course of historical development, a course that would eventually lead through capitalism to an Indian transition to socialism (Wolfe, 1997: 390).

There was a diffusion of the colonial attitude of viewing Western culture as 'superior' to traditional culture. This, for instance, had significant impacts on education and the medium of instruction in the schools and is discussed elsewhere in this note.

Today, the countries in the region are welfare states performing a range of welfare functions. Education, health, law and order, employment, infrastructure and public utility services, to list a few, are important functions that the state carries. States have a huge role in development which differs qualitatively from that in developed countries. The welfare component is explicitly set out in the provisions enshrined in the constitution of each country. Planning, policy making, budgetary allocations and actual provision of services rest with the State. For example, in Maldives, city and island councils' responsibilities include providing roads, waste disposal, pest control, water, electricity and sewage systems along with primary healthcare, preschool education and educational and vocational programmes for adults. (http://www.clgf.org.uk/maldives/. Accessed on December 9, 2015).

Challenges of Linguistic Diversity

Language has a great functional value as it is used for administrative and communication purposes and influences educational outcomes. South Asia is an area of staggering linguistic diversity and the socio-political challenges that such diversity presents are innumerable; these challenges have to be spelt out to understand the implications for Education for Sustainable Development. India has twenty-two official languages, including English, each language, other than Sanskrit, spoken by more than one million people. Sri Lanka (Sinhala and Tamil) and Pakistan (Urdu and English) have two official languages each.

The first and foremost challenge that diversity poses is in the language chosen for the provision of education. "While uniform solutions for plural societies, may be both administratively and managerially simpler, they disregard the risks involved both in terms of learning achievement and loss of linguistic and cultural diversity" (Nakashima, Prott and Bridgewater, 2000:12). Education is inherently normative in nature – what is taught, what is learnt, the language in which it is transacted, the goals and systems foster normativeness. Formal "…education systems have disrupted the practical everyday life aspects of indigenous knowledge and ways of learning, replacing them with abstract knowledge and academic ways of learning. Today, there is a grave risk that much indigenous knowledge is being lost and, along with it, valuable knowledge about ways of living sustainably" (http://www.unesco.org/education/tlsf/mods/theme_c/mod11.html. Accessed on May 6, 2013).

By not using mother tongue for instruction, language as the tool for thought has become language for rote-learning leading to a loss in deep learning. Opportunities for discourse-based learning, with members of family, community and peers, have been wiped away. This has implications on cognitive development leading to low esteem of the learners. Deepa, in her ninth article in the multi-part series *Lens on Education*, shares the observation made by the Nepalese scholar Malla on the high dropout rates in higher secondary schools where students simply dropped out of school given the prospect of learning in English (Deepa, 2006).

Language and Education

Education systems in the region were either under the influence of colonial rule or have been under the influence of strong neighbours like India which itself was under colonial rule. Due to the region's colonial history, English has been in use in education and communication and has a wide acceptance in the region, as it emerged over time as the language of mobility and gainful employment. People aspire to be educated in English referring to the use of English as a medium of instruction and consequently the vehicle for transmission of knowledge. This is distinct from learning a language, whether mother tongue or not, with its vocabulary, grammar, speaking, reading and writing. English was subsequently chosen by South Asian countries once they acquired sovereignty because of its non-nativeness and the neutrality it stood for. However, such dispensation impacts on the transmission of sophisticated knowledge and wisdom held by communities.

The predisposition for English as a medium of instruction has led to neglect of local languages and mother tongues to a point where these have lost out on quality aspects. Paradoxically, speakers of mother tongues face a double disadvantage from receiving instruction in a foreign language be it national/local language or English. Both knowledge and opportunities are available to those who know the foreign language: English. While there are strong educational arguments in favour of mother tongue instruction, a careful balance also needs to be made between enabling people to use local languages in learning and providing access to global languages of communication through education (UNESCO, 2003: 1).

Topography and Climate

South Asia comprises of three per cent of the world's land mass and contains a markedly varied topography. The topography and the climate have shaped and supported an exceptional array of biodiversity. South Asia contains diverse ecosystems, such as the mountains of the Himalayan Hindu Kush, the rangelands of Bhutan, the Thar Desert, the high altitude freshwater lakes of Nepal, the Deosai plains in Kashmir, the extended contiguous mangrove swamps of India and Bangladesh and the coral reefs and atolls of Maldives. "Forests range from tropical, subtropical, and coastal to temperate, and the deserts range from hot to cold. These ecosystems occupy about 3.6 percent of the world's area but contain 16 percent of floral and 12 percent of faunal species found in the world" (http://siteresources.worldbank.org/SOUTHASIAEXT/Resources/Publications/448813-1231439344179/5726136-1259944769176/SARclimagechangechapter11november2009.pdf. Accessed on December 9, 2015). "The region is ecologically diverse and houses two of the thirty-four global biodiversity hotspots, namely the Western Ghats and Sri Lanka Hotspot and the Himalayan Hotspot. The Himalayan Hotspot extends over 3000 km through northern Pakistan, Nepal, Bhutan and the north-western and north-eastern states of India and includes all the world's high mountain peaks over 8000 m" (http://www.conservation.org/where/priority_areas/hotspots/asia-pacific/Himalaya/Pages/default.aspx. Accessed on May 6, 2013).

India is one of the world's twelve mega-biodiversity centres, and the subcontinent is one of the six Vavilovian centres of origin of species (http://en.wikipedia.org/wiki/Center_of_origin. Accessed on May 6, 2013).

Biomass-Based Agrarian Region

South Asia is essentially a biomass-based agrarian region, directly attributed to the ecological diversity of the region. Much of the population directly depends on natural resources for their living. Whilst industrialization is taking place particularly in India, the six countries are, despite trends towards urbanization, predominantly rural with between 63 % (Bhutan) and 83 % (Nepal) living outside urban areas and relying on a rural economy. On average about "three-quarters of South Asia's population lives in rural areas, with one-third living in extreme poverty (on less than a dollar a day)" (Singh and Gonzalez, 2013, p.69). Agricultural production as a percentage of total GDP ranges from 4 % in the Maldives to 34 % in Nepal around a mean of 16 % for the six countries as a whole. Income disparity divides peoples within each of the countries and their capacity to act for a sustainable environment. Historically, the region has been one of the least developed in the world, and the challenges like climate change, degrading resource base and poor educational levels are not quite so favourable for redeeming the region from the demographic and geographic situation.

Population Pressures

Demand on these ecosystems is high. Though South Asia occupies only 4.8 % of the world's land mass, it is home to about 20 % of the world's population. This figure is expected to rise to about 25 per cent by 2025 (UN, 2015: 11).Within this, India is far and away the most heavily populated at 1,311.1 million while the Maldives contains some 360,000 people and Bhutan has a population of 770,000. Between these extremes are Bangladesh (160.99 million), Nepal (28.51 million) and Sri Lanka (20.71 million). In South Asia as a whole, the annual percentage growth rate has slackened from 2.15 % in 1990 to 1.5 % but this has to seen in the light of a current estimated global growth rate of just below 1.1 %. Population densities are so high that the pressure on land is a commonly recurring feature.

Human development, which is about "creating an environment in which people can develop their full potential and lead productive, creative lives in accord with their needs and interests" (UNDP, 2001: 9), is being adversely affected by the growing population. Interestingly, despite population being seen as a problem, currently the region is trying to espouse the "demographic dividend" hypothesis which says that a huge population, if well managed, will be a positive contribution to sustainable and inclusive growth rather than a liability.

Environmental Challenges

Environmental issues are omnipresent in South Asia and inflict tremendous environmental damage. The capacity to withstand stress stands severely endangered as of now (UNEP, 2008: xii). Some of the priority issues of the region are climate security, food security, water security, energy security and urbanization. South Asia faces problems with respect land, water, air, flora and fauna. These are inter-related and driven by natural hazards on the one hand and the human condition – population growth, urbanization, poverty, failures in institutions and governance, unfettered development and globalization of trade – on the other.

Environmental risk is a very complicated phenomenon that needs to be tackled through a mix of policies and incentives. Context determined economic, environmental and social policies, supported by education, offer the possibility of overcoming the challenges to achieve durable and inclusive growth. There does not seem a lack of interested bodies and the need for improvements in monitoring regulation and work on the ground is recognized as vital but this is often hampered by institutional constraints and limited infrastructure. The countries have had regulations pertinent to forestry since the mid-19th century colonial period. Following the UN Stockholm conference in 1972, the countries enacted more broadly based environmental legislation. However, Schukoske (1996) has noted with reference to Sri Lanka, but it is probably equally true in the other countries, that:

> Due to the tension between environmental protection and economic development…enforcement of environmental laws presents special challenges. Changes in social attitude about use of the environment come slowly and sometimes grudgingly. Enforcement of environmental laws is often at odds with traditional uses of natural resources by indigenous peoples and with income producing uses by owners and investors (p. 155).

Vulnerability to Climate Change

South Asia is very vulnerable to the effects of climate change. Impacts of climate change have been observed in the form of glacier retreat in the Himalayan region; it is estimated that the approximately 15,000 glaciers will likely shrink from the present total area of 500,000–100,000 km^2 by 2035 with all the consequences that this will entail (UNEP, 2008: xiv). At the other extreme, the low lying atolls of the Maldives seem to be facing an increasing frequency of inundation.

Climate change and biodiversity are closely linked and each impacts the other. Biodiversity is threatened by human-induced climate change, but biodiversity reduces the impact of climate change. The presence of healthy biodiversity builds natural resilience to climate extremes: as an example, forests are nature's social security check in times of disaster and crisis; additionally forests also act as a sink for harmful greenhouse gas emissions.

Changes in climate are likely to affect biodiversity directly and indirectly. Direct effects are mostly related to temperature and precipitation changes that affect species distributions and ecosystem compositions. Species to move to higher elevations in search of more suitable habitats and may increase the risk of extinction for already vulnerable species with limited climatic ranges and restricted habitats.

Biodiversity changes are also occurring as a result of habitat destruction. Forests account for about 20–30 % of the total land area of India, Nepal and Sri Lanka and about 68 % in Bhutan. (http://web.worldbank.org/WBSITE/EXTERNAL/COUNTRIES/SOUTHASIAEXT/0,contentMDK:22405929~pagePK:2865106~piPK:2865128~theSitePK:223547,00.html. Accessed on December 9, 2015). Savannas and dry forests are grazing areas for large herds of livestock, which are essential elements of food supply and rural transport. Climate change will affect the vegetation, productivity and biodiversity of these ecosystems.

Climate change has also impacted the coastal environment. Coral coverage in the Indian Ocean islands and South Asia combined has declined from more than 40 % in 1997 to just over 20 % in 2002 (http://web.worldbank.org/WBSITE/EXTERNAL/COUNTRIES/SOUTHASIAEXT/0,contentMDK:22405929~pagePK:2865106~piPK:2865128~theSitePK:223547,00.html. Accessed on December 9, 2015).This is translating into losses to fisheries, mangrove ecosystem productivity and tourism. Coral reefs are not unlike tropical rainforests in biodiversity and are important sources of tourism revenue. Along with mangroves they also protect coastal areas against beach and cliff erosion, sedimentation, floods and storm surges. The irreversible losses to biodiversity adversely impacts longer-term economic opportunities in coastal communities.

The degradation of water quality with attendant health problem from water borne diseases is an issue across the region as is water scarcity particularly in the drier north-western parts of the region. Except for Bhutan and Nepal, the per capita water availability in the region is less than the world average, with water use in this region being limited mainly to the agriculture sector.

References

Climate change: Are south Asia's ecosystems at the brink of extinction? http://web.worldbank.org/WBSITE/EXTERNAL/COUNTRIES/SOUTHASIAEXT/0,,contentMDK:22405929~pagePK:2865106~piPK:2865128~theSitePK:223547,00.html. Accessed 9 Dec 2015.

Colonialism in South Asia. The University of Texas at Austin, blog by Christopher Rose, 2013. http://blogs.utexas.edu/15minutehistory/2013/01/23/episode-9-the-end-of-colonialism-in-south-asia/. Accessed 8 May 2013.

Commonwealth Local Government Forum. http://www.clgf.org.uk/maldives/. Accessed 9 Dec 2015.

Conservation International. *Himalaya*. http://www.conservation.org/where/priority_areas/hotspots/asia-pacific/Himalaya/Pages/default.aspx. Accessed 6 May 2013.

Deepa, A. (2006). *Mother tongue or English? Lens on education*. No. 9. India: India together. Ecosystems and biodiversity, Chapter 11 http://siteresources.worldbank.org/SOUTHASIAEXT/Resources/Publications/448813-1231439344179/5726136-1259944769176/SARclimagechangechapter11november2009.pdf. Accessed 9 Dec 2015.

History of Bhutan. http://cypresshimalaya.com/history-of-bhutan/. Accessed on 9 Dec 2015.

History of British Ceylon. https://en.wikipedia.org/wiki/History_of_British_Ceylon. Accessed 9 Dec 2015.

Maldives History. http://www.globalsecurity.org/military/world/indian-ocean/mv-history.htm. Accessed 9 Dec 2015.

Nakashima, D., Prott, L., & Bridgewater, P. (2000). *Tapping into the world's wisdom. UNESCO sources*. No. 125. (p. 12). Paris: UNESCO.

Sarabhai, K. (2000). *Securing our future in the new century: Lessons from India*. New Delhi: IUCN. http://www.ceeindia.org/newintra/papers/Securing%20our%20Future.htm. Accessed 6 May 2013.

Schukoske, J. E. (1996). Enforcing environmental laws in Sri Lanka through fundamental rights litigation. *International Legal Perspectives, 8*, 155.

Singh, A., & Gonzalez, E. (2013). *Millennium development goals and community initiatives in the Asia Pacific*. New Delhi: Springer.

United Nations. (2015). *Demographic and social statistics*. http://unstats.un.org/unsd/demographic/default.htm. Accessed 10 Aug 2015.

United Nations Development Program. (2001). *Human development report 2001*. New York: UNDP. http://hdr.undp.org/sites/default/files/reports/262/hdr_2001_en.pdf. Accessed 11 Dec 2015.

United Nations Educational, Scientific and Cultural Organization. *Indigenous knowledge and sustainability*. http://www.unesco.org/education/tlsf/mods/theme_c/mod11.html. Accessed 6 May 2013.

United Nations Educational, Scientific and Cultural Organization. (2003). *Education in a multilingual world*. Paris: UNESCO.

United Nations Environment Programme and Development Alternatives. (2008). *South Asia environment outlook 2009: UNEP, SAARC and DA*. Nairobi: UNEP.

Wikipedia. (2014a). *Center of origin*. http://en.wikipedia.org/wiki/Center_of_origin. Accessed 6 May 2013.

Wikipedia. (2014b). *Indian subcontinent*. http://en.wikipedia.org/wiki/Indian_subcontinent. Accessed 6 May 2013.

Wolfe, P. (1997). History and imperialism: A century of theory, from marx to post-colonialism. *The American Historical Review, 102*, 390.

Chapter 1
Schooling for Sustainable Development in South Asian Countries

Rajeswari Namagiri Gorana and Preeti Rawat Kanaujia

1.1 Introduction

The success of Environment and Development Summits (United Nations Conference on the Human Environment in Stockholm, Sweden, 1972, United Nations Conference on Environment and Development in Rio de Janeiro, Brazil, 1992, United Nations Commission on Sustainable Development Johannesburg Summit, Johannesburg, South Africa, 2002) led to the emergence of the catchphrase 'sustainable development.' "The term suggested promise, promoting a framework that would integrate the four entities—society, environment, government, and business, in a common process of development, focusing on the present but respecting the needs of future generations as well as the memories and desires of past ones" (Fergus and Rowney 2005: 60).

By conjoining development and sustainability, development has acquired a new paradigm—a paradigm with a philosophical base for balancing and integrating environmental aspects, economic concerns and societal advancement. In South Asia, one hears about sustainable development in the media, academic discourses, education policies and government initiatives and, seemingly, the term has had a profound impact on influential groups, organizations and active individuals. It can be argued that paradoxically it is not visions of development but issues of

R.N. Gorana (✉)
Programme Coordinator, Children's Media Unit, Centre for Environment Education (CEE), Ahmedabad, India
e-mail: rajeswari.namagiri@ceeindia.org

P.R. Kanaujia
Senior Programme Coordinator, Northern Regional Cell, Centre for Environment Education (CEE), Lucknow, India
e-mail: preeti.rawat@ceeindia.org

R.N. Gorana, P.R. Kanaujia (eds.), *Reorienting Educational Efforts for Sustainable Development*, Schooling for Sustainable Development,
DOI 10.1007/978-94-017-7622-6_1

unsustainable development—underdevelopment as well as haphazard development— are the drivers of Sustainable Development in the region.

Education received a stimulus from the environment and development summits and widened its agenda to include sustainability. In the South Asia region, education has been a means of addressing the challenges of nation building and social issues like discrimination, and inequality. In recent times, it emerged as a staple response to issues like resource degradation, livelihoods, health, climate change, peace, disaster risk reduction and preparedness—probably such a dispensation of education in a true sense has grasped the sophisticated notion of sustainability. "The concept of sustainability encompasses not only environment but also poverty, population, health, food security, democracy, human rights and peace. Sustainability is, in the final analysis, a moral and ethical imperative in which cultural diversity and traditional knowledge need to be respected" (UNESCO 1997: 2).

United Nations organizations like UNESCO and programmes like the United Nations Environment Programme (UNEP) and the United Nations Development Programme (UNDP), the Earth Summits, international treaties and agreements, international funding agencies and governments provided a powerful focus for education policy makers and practitioners to work with sustainable development, lending it a wider focus. Initially, the International Environmental Education Programme (IEEP) of UNESCO and the 1977 Intergovernmental Conference on Environmental Education in Tbilisi introduced Environmental Education (EE) into formal education. "The Thessaloniki Declaration tried to unite two contesting discourses: the discourse of environmental education, especially as this had been defined in Tbilisi, in 1977 and the discourse of sustainable development, as was defined in the Brundtland Report, *Our Common Future* (The World Commission on Environment and Development 1987)…The Declaration proposed, in a sense, a new definition for sustainable development by linking it directly with education" (Nikolopoulou et al. 2010: xiii). The 'Ahmedabad Declaration 2007: A Call to Action', the fourth in the series of Conferences on Environmental Education held at Ahmedabad in 2007 (following on from the first Intergovernmental Conference held in 1977 at Tbilisi, Georgia) proclaimed that "…through education we can learn to prevent and resolve conflicts, respect cultural diversity, create a caring society and live in peace. We can learn from indigenous and traditional patterns of living that respect and honour the Earth and its life-support systems and we can adapt this wisdom to our fast-changing world" (http://aries.mq.edu.au/pdf/AhmedabadDeclaration.pdf. Accessed on November 24, 2015). These developments have laid the foundation for Education for Sustainable Development (ESD). ESD encourages a shift from viewing education as a delivery mechanism to a lifelong, holistic and inclusive process. Significantly, it recognizes the role of education as an effective driver of change. The focus clearly is on the learner and learning—learning for sustainability in its broader sense. With this, the focus shifted to contexts—cultural, social, historical, linguistic, and ecological—of the learners and the learning.

The positive links between schooling and development are well recognized universally. Education, in the South Asian region, is aspirational. At the level of individual people, education is seen as the fastest way to a better future—a future that could not be achieved by the earlier generations. Individuals turn to education with the belief that it would offer improved life circumstances and employment. "Aspirations do not exist within a vacuum, but rather occur within a social context or against a changing social context" (Gutman and Akerman 2008: 3). Economists have long recognized the importance of human capital in economic growth. Governments have seen education as an effective driver to comprehend, assimilate and infuse nationalist ideology for nation building. For example, the National Educational Planning Commission (NEPC) of Nepal was explicit when it stated that "...schools and educational systems exist solely for the purpose of helping the youth of a nation to become better integrated into the society" (NEPC 1956: 14). Education became a tool for affirmative action and to address inequality and discrimination. Education is seen as central to social and economic mobility. With sustainable development co-opting education, the economic context to learning will have to be reexamined. The United Nations declared the beginning of the Decade of Education for Sustainable Development (DESD 2005–2014).

> The basic vision of the DESD is a world where everyone has the opportunity to benefit from education and learn the values, behaviour and lifestyles required for a sustainable future and for positive societal transformation (UNESCO 2006: 2).

1.2 Education in South Asia

Teaching and learning are not new activities in the region. Formal and informal learning were practised traditionally through transmission approaches to learn religion, sciences, art and craft with the focus clearly on economics, culture and religion. It was not an organized activity on an equal footing, and was exclusionary, since it was not accessible to such groups as the lower castes and women. Much of it was oral. Teaching and learning as we experience today demonstrate a great transformation as contexts and needs have changed.

Mass education is, evidently, new to the countries of the region and has been prioritized as a means to facilitate modernization and economic growth under state sponsorship after the countries in the region gained their sovereignty. Before its advent as a modern practice in the form of mass schooling, education was provided by religious bodies. In the Maldives, schools, privately owned or operated by the island communities, taught children to read the Quran, and to read and write in Dhivehi, the mother tongue of Maldivians. Traditionally, literate monks in Sri Lanka taught the Sinhala alphabet and elementary Buddhist literature to privileged students in temple schools. In Nepal, "there were basically two types of traditional and indigenous practices: individual efforts and family tradition of educating the children, and religious nature of educational institutions. Education based on Hindu tradition (particularly Gurukul), Buddhist education at Gumbas (Monasteries) and Muslim education at Madarsas were providing religious education on a limited basis" (Status of Education in Nepal and Development Plan, Gangalal Tuladhar (http://web.isc.ehime-u.ac.jp/ice/6-1@Gangalal%20Tuladhar_text%20_8p.pdf).

Accessed on May 7, 2013). Formal education in India was traditionally based on literature and scriptures such as the Vedas and other Hindu texts, and was available for the privileged Brahman class and the rulers. Both formal and informal ways of learning were prevalent for art, craft, and music. Education was based on practical needs alongside religious education. Educational efforts within the traditional communities in India were concerned with "the transmission of various skills, especially those related to the economic life of the community: agriculture, hunting, fishing, and caring for its environment—its trees, animals, birds, water bodies, etc" (NCERT 2006: 2).

Contemporary school education or formal education has roots in the colonial times of the region. Colonization changed the face of education, in countries like India and Sri Lanka.

> During the sixteenth century, Portuguese missionaries established schools for promoting Roman Catholic culture. In 1656, as Dutch came in, a well-organized system of primary schools was set up. During British period, government-funded schools and Christian schools began to expand. In 1870, a series of events revolutionized the education system in Sri Lanka. As part of a policy to promote universal literacy, education became free in government schools, elementary and technical schools were set up in rural areas, and vernacular education received official encouragement. When independence came in 1948, Sri Lanka had a well developed education infrastructure. Although still hampered by gross ethnic, geographic, and gender inequalities, it formed the basis for a modern system (Ross and Savada 1988).

In British India, the East India Company and the British Crown,

> "created a new state system of education that largely replaced the former indigenous network of schools….The initial policies introduced by the directors of the East India Company promoted an elitist English education geared towards a small minority of the population—the higher castes and classes (The East India Company controlled the Indian sub-continent until the Revolt of 1857. After the Revolt, the British Crown took over control from the Company)". An important goal of these policies was to "form a class of people who may be interpreters between us [the British] and the millions whom we govern, a class of persons Indian in blood and color, but English in taste, in opinion, in morals and intellect" (Macaulay's Minute on education in 1835 as quoted by Chaudhary 2007).

There is much to be satisfied with education in the region. There have been considerable achievements in terms of achieving literacy levels; bringing children, particularly girls, to school and retaining them; attaining educational levels that have helped the nations of the region to make a mark in various fields; address social ills; and influencing children's and women's health and population growth. Education, through formal, non-formal and informal avenues, offered by the states and NGOs, various trusts and boards, has moved beyond the basic education or literacy goals to focus on livelihoods, self-sufficiency, entrepreneurship, conservation, health and hygiene, children's and women's welfare.

Though the larger benefits of education are strongly recognized in the region, there is a tremendous heterogeneity of educational achievements in the countries of the region. The achievements and gaps in education have implications for EE and ESD in the region as it is the existing educational systems and structures that will have to integrate EE and ESD qualitatively and quantitatively.

1.3 Understanding Environment and Development in South Asian Contexts

The Stockholm Conference has a close link to South Asia. It was where Mrs. Indira Gandhi, the then Prime Minister of India, drew the seminal linkage between environment and development by proclaiming insightfully that "…the environment cannot be improved in conditions of poverty. Nor can poverty be eradicated without the use of Science and Technology". She stressed that, "…life is one and the world is one, and all the questions are interlinked. The population explosion, poverty, ignorance and diseases, the pollution of our surroundings, the stockpiling of nuclear weapons and biological and chemical agents of destruction are all but parts of a vicious circle. Each is important and urgent but dealing with them one by one would be a wasted effort…" (Roy and Sharma 1991: p. 29).

Some 40 years later, The 2011 Human Development Report (*Sustainability and Equity: A Better Future for All*) argued that the urgent global challenges of sustainability and equity must be addressed together. Bold action was needed on both fronts, the report contends. It identified policies on the national and global level that could spur mutually reinforcing progress towards these interlinked goals. It notes that the recent human development progress for most of the world's poor majority must be sustained, for the benefit of future generations as well as for those living today. Further, the report shows how the world's most disadvantaged people suffer the most from environmental degradation, including in their immediate personal environment, and disproportionately lack political power, making it all the harder for the world community to reach agreement on needed global policy changes. It also outlines great potential for positive synergies in the quest for greater equality and sustainability, especially at the national level. Further it emphasizes the human right to a healthy environment, the importance of integrating social equity into environmental policies, and the critical importance of public participation and official accountability (http://hdr.undp.org/en/content/human-development-report-2011. Accessed on May 6, 2013).

The first and foremost point to take note of is that any education or development intervention has to be tuned to the specific conditions of the countries in South Asia. In this context, Kartikeya Sarabhai in his article on Leapfrogging Development makes some pertinent observation with respect to the UNEP campaign slogan for World Environment Day 2008 'CO_2: Kick the Habit!' What would such a slogan mean to a person in a village in India who perhaps has never had an electricity connection, has no running water and no toilet? Which CO_2 intensive activity was a poor rural Indian family meant to "kick"? Or take those who live in a city in India. Forty percent of urban dwellers in India live in slums with no real access to energy intensive amenities. How are they to kick a habit that they could never afford to have in the first place? It should be noted that a new slogan, "CO_2: Pick Right", was coined for India to transform the campaign to urge people to consider the need to make developmental choices that are sustainable (Sarabhai 2010: 69).

Development has to be seen in its socio-historical context. The decolonization of South Asia had begun at the same time as World War II ended and countries of the region became sovereign states. Governments set about rebuilding their nations the way they envisioned for their people. The approach to development continued to be informed by colonial thought or, put another way, on a Western model of development. "Placed in this context, development simply became another name for economic growth. The rationale was that economic growth should be paramount. Economic growth would alleviate poverty by creating wealth, which could then be used to solve 'social' problems. This separation of the increasingly economic realm began to define social and cultural aspects for Third World populations" (Banerjee 1999: 5).

The second important point is that any intervention cannot be imitative. In 1992, the India report to United Nations Conference on Environment and Development (UNCED) at Rio stressed that the "real challenge of development was not how to get there but how not to." The definition of development came from "strong commercial forces and a media that defines development and progress as a process of westernization." The observation made in the report was an indication of the need to find alternative development models, and a caution not to see development as a 'catching-up development.' The unsustainability of predominant Western models of development based on consumption were already being felt and acknowledged by the world community. The report stated, "You cannot have sustainable development towards an unsustainable goal" (Sarabhai 2000: 8).

> Education, as a driver of change has a bigger and context driven agenda for countries of South Asia when consciousness for environment and development poses both a challenge and opportunity. The countries, "needs to move forward, but directly, to a sustainable state rather than first imitate lifestyles and resource use models as they exist in developed countries and only then go towards sustainability. It is this challenge of 'leapfrogging', of having the confidence to make critical choices and walk on paths which have never been traveled before that defines the goals of ESD" (Sarabhai 2009: 3).

1.4 Environmental Education and Education for Sustainable Development

The case for taking up, promotion and development of EE in South Asia did not arise from the predominant western notion of outdoor, nature, or conservation education. Though its formal beginnings can be traced to Recommendation 96, made at the United Nations Conference on the Human Environment in Stockholm in 1972, the Recommendation calls for the provision of environmental education as a means to address environmental issues worldwide and the establishment of an international programme in environmental education.

> Environmental Education and care is not new in the region. Centuries of frugal and careful tending of the land on which communities have directly depended has resulted in living in harmony with nature. Each generation has passed on its profound understanding of how to

> manage traditional natural resources. Religious philosophy has a bearing on the traditional relationship with environment in many nations. In Bhutan, for example, the Buddhist Philosophy is seen as a major reason for environmental protection and prevention of environmental disturbance (Saeed et al. 1998: 18).

EE was a response to the strong need to address socio-economic aspects linked with environment and development, thus, by implication, expounding on the notion of education for sustainable development much before the Thessaloniki conference in Greece in 1997. At Thessaloniki, environmental education was explained as an education for sustainability. At this point the international community recognized the relationship of sustainable development and environmental education.

Education was seen as "…. a indispensable means to give all women and men in the world the capacity to own their lives, to exercise personal choice and responsibility, to learn throughout life without frontiers, be they geographical, political, cultural, religious, linguistic, or gender" (Nikolopoulou et al. 2010: xii). A complex interplay of economic, social and cultural trends has created a landscape for making EE and ESD in South Asia an empowering practice.

Internationally, debates sought to clarify EE and ESD. Hesselink et al. (2000: 12) claimed however that, "ESD differs significantly from the naturalist, apolitical and scientific work carried out under the EE banner in the 1980s and early 1990s. Now EE practices are closing the gap as they focus more sharply on developing closer links between environmental quality, ecology, and socio-economic and the political threads that underlie these. ESD has a strong futures component. It involves learners in an examination of probable and a possible future…this exercise is crucially linked to the development of 'hope' empowerment and action."

Environmental education came into general use through the efforts and initiatives of various actors in the region. The distinctive philosophy, approaches, methods and implications of EE and ESD can be attributed to the use of education as an intervening tool to address development challenges. There is a marked imprecision in the way EE and ESD are understood and defined. This can be attributed to many reasons, chief among them being, the surfeit of adjectival education initiatives (water and sanitation education, peace education, population education, disaster risk reduction education, climate change education) and innovations that have been dovetailed with formal education, particularly school education. It has led to a fragmented approach to EE and subsequently to ESD, albeit that the fragments themselves are parts of a whole. Further, Abe and Bhandari in 1991 observed that, "the awareness about Education for Sustainable Development (ESD) is quite low in the region. EE is understood in the domain of formal education, both at the secondary and higher levels. The view that ESD is an extension of EE is prevalent among practitioners, educators and government officials" (UNESCO 2005: 55).

The terms EE and ESD are either used interchangeably or together which the authors continue to do so here in this chapter. While there are country specific efforts to promote EE and ESD, there are also region-wide efforts to span South Asia, South and South East Asia and the Asia Pacific. These are facilitated by intergovernmental organizations. For example, the South Asia Co-operative Environment Programme (SACEP) has, since 1982, the following mandate:

> The Mission of SACEP is to promote regional co-operation in South Asia in the field of environment, both natural and human in the context of sustainable development and on issues of economic and social development which also impinge on the environment and vice versa; to support conservation and management of natural resources of the region and to work closely with all national, regional, and international institutions, governmental and nongovernmental, as well as experts and groups engaged in such co-operation and conservation efforts (http://www.sacep.org/html/about_visionmission.htm. Accessed on May 6, 2013).

"The South and Southeast Asia Network for Environmental Education (SASEANEE), is a network of agencies and individuals involved in or interested in networking, initiating, or supporting environmental education programmes in the region" (http://www.saseanee.org/. Accessed on May 6, 2013). UN agencies have been important in the EE and ESD initaitves in many countries of the region. IUCN, known as the World Conservation Union founded in 1948, has played a critical role in the development of EE in the region in the fashion noted by Palmer that "organizations (like IUCN) concerned with the development of environmental education moved towards defining its meaning and promoting its legitimacy." (Palmer 2002: 5). IUCN, the world's first global environmental organization strived to mainstream EE and ESD in the region. The Institute for Global Environmental Strategies or IGES conceived the Environmental Education project under which the Japanese government paid serious attention to developing schemes for Environmental Education in Japan and Asia (IGES 1999: 9).

These initiatives have had an impact, and Abe and Bhandari (2001) note that

> Preliminary analysis reveals that people within the Asia-Pacific region are conscious of, and responsive to, environmental problems and their emerging consequences. They are deeply concerned with the issue of environmental deterioration. The notion that we must protect the environment and protect humanity is growing in the region.
>
> A high level of awareness is demonstrated by the:
>
> 1. incorporation of environmental concerns into curricula and educational programs,
> 2. initiation of eco-business activities,
> 3. involvement of students in extra-curricular activities,
> 4. active involvement of non-governmental organizations (NGOs) in raising awareness, and
> 5. determination and commitment of government. At the national level, governments have shown their interest by emphasizing the urgency of incorporating environmental concerns into formal and non-formal education systems.

Formal, non-formal and informal approaches of education are used for EE and ESD. They are used to provide and encourage awareness, knowledge, training and skill building, attitudinal change and opportunities to act. Non-formal education takes place both within and outside educational institutions and caters to all ages. Depending on country contexts, it may cover educational programmes for adult literacy, basic education for out-of-school children, life-skills, work-skills, and general culture. Non-formal education programmes do not necessarily follow the 'ladder' system, may have varying duration, and may or may not confer certification of the learning achieved.

Significantly, again as noted by Abe and Bhandari (2001), NGOs have actively promoted EE, widened its scope and practice, and importantly have kept up the momentum.

> Major NGO activities include: (1) producing environmental awareness materials and conducting training, camps, field activities and research programs; (2) mobilizing community resources and linking income-generating activities with EE; (3) guiding conservation activities for environmental clubs; (4) providing press releases to raise public consciousness; (5) establishing environmental networks such as the Green Volunteer Network and Green Bazaar; (6) implementing innovative activities such as demonstration projects, interpretation centers and alternate models of education; and (7) lobbying the media, government, industry and business for viable education programs (Abe and Bhandari 2001: Executive Summary, p.X).

Notably, Savalia (2011) notes that though it is rural communities which form a considerable part of the population in developing countries and are key catalysts of change in achieving ESD, they have not been engaged sufficiently well. He points out that "Existing institutional mechanisms for ESD have been largely unable to engage in non-formal and in-formal education in comparison to their involvement with formal education. So there is a need for more focus on NFE as an ESD. NFE has the added advantages of belonging to non-rigid systems, remains locality specific, develops in a need based manner and can develop and evolve from a participatory leaning for SD" (http://www.accu.or.jp/esd/forum_esd_2010/program/program12_01/pdf/text2.pdf. Accessed on May 7, 2013).

Almost all the countries in the region have established free and compulsory education as recommended in Article 26 of the Universal Declaration of Human Rights. However, in Article 26, only school education is included (Kazuo 2011) (http://www.unescobkk.org/education/news/article/reforming-international-cooperation-in-education/. Accessed on May 7, 2013). and important aspect which the Article does not mention is informal education. Education occurs in the family and community albeit in an unorganized way; traditionally, education took place outside schools through these venues. It was a part of life and living. That is why learning itself is a human right because without learning a human being cannot be a human being.

Educational institutions formalize and systematize learning. In the region, some level of formalization can be seen with respect to EE and ESD. Because of the structural characteristics of and inertia within, formal education systems, EE initiatives have been taken up as extra- and co-curricular activities before and even after being mandated to be part of the curriculum (EFA 2015, http://unesdoc.unesco.org/images/0023/002305/230507e.pdf. Accessed on December 8, 2015). "Eco-clubs, nature camps and Environmental Brigades are gaining popularity in South Asia, and are catalysts for imparting awareness, knowledge, attitudes and skills among children and youth" (UNESCO 2005: 56).

General trends and patterns of formal EE in the Asia and Pacific region indicate that there is a tendency towards the:

1. incorporation of EE into all forms of education;
2. writing of greener curricula;
3. creation of new initiatives such as green accounting, eco-clubs, eco-farming, green press and green banking;
4. perception of EE as a new approach to education;
5. development of composite courses at the primary level;
6. expansion of focus from physical science to social science courses as well;
7. increase of emphasis on formal education;
8. establishment of successful eco-business activities;
9. emphasis on extra-curricular and co-curricular activities to supplement school curricula; and
10. professional development of environmental educators and facilitators. (Abe and Bhandari 2001: Executive summary, page ix).

Within formal education, EE has taken the form of 'integration, infusion, separate subject and project based learning' (Abe and Bhandari 2001). 'Learning about the environment; Learning through the environment; and Learning for the environment' is the underpinning strategy of EE in formal education. "Environment is a medium of education…used as a source of stimulation for realistic activities with development of skills and attitudes as an important purpose. It is known as environmental studies" (Palmer 2002: 9). Environmental Studies here reflects an *integration* approach—wherein both social and environmental aspects covered by the social studies and science subjects become the medium of learning. This integrative approach is usually seen in use at the primary level (up till standard V). See India chapter on Textbooks and Bangladesh chapter. *Infusion* is the approach followed for secondary levels (up till standard X), which allows sufficient environmental perspectives to be added to science and social studies. India introduced project based learning for senior secondary levels (Standards XI and XII).

There are varied examples in the region on how curriculum in formal schooling is being dealt to bring in EE concepts. "Cross-curriculum approaches have been adopted to integrate environmental themes into curricula. In this approach, the whole curriculum is reviewed and environmental concerns are incorporated into all subjects, not just one particular subject. Some countries have begun to "green" their curricula by incorporating environmental concerns and have emphasized the use of local resources in teaching and learning processes. This involves the integration of environmental principles, problems and solutions into other disciplines. Both the natural environment and the man-made environment are involved" (Abe and Bhandari 2000: 63).

UNESCO-UNEP's EE newsletter, 'Connect' which focused on the aspects related to developing an Environmental Education Curriculum states…choice of the infusion method as better than that of a separate subject for transposition of the environmental knowledge, cognitive skills and attitudes acquired in the classroom to the decision-making processes of the learners throughout their lives…experience and research have demonstrated that EE must be taught—through problem solving in a variety of different, even confusing, situations to take into account environmental consequences in their economic courses, social studies and other disciplines (UNESCO-UNEP 1989).

In India, with considerable experimentation in mainstreaming EE in to the school system, the *infusion* and *separate subject* approaches have attracted much debate and discussion. There is a strong resistance to make EE a separate subject in India. It is argued as neither desirable nor possible due to logistical challenges. Perhaps, a strong philosophical and pedagogical reasoning for taking the infusion approach came from Indira Gandhi.

> I do not suggest that we should develop newly structured courses and curricula, but that we should introduce a new orientation to learning processes and attitudes. For example, an awareness of environment and the need for conserving basic life support systems like soil, water, flora and fauna, the need for population stabilization and family planning etc. should grow naturally into the students from the young age. Since opinions are usually formed in the formative years of one's life, teaching aids, methods and open door field visits should be so structured that living in harmony with the environment becomes a habit with each young individual (as quoted in Bandhu et al. (1989). https://environmentalartilces.wordpress.com/indira-gandhiji-1917-%E2%80%93-1984-environmentalist/. Accessed on December 8, 2015).

"In the report prepared for Ministry of Environment and Forests in May, 1998, CEE had recommended that the preference for the infusion approach is because of the strong argument that environment is not a separate compartment or subject—in fact it infuses every aspect of life and every subject must be taught environmentally so that students get a holistic perspective…EE is not merely an add on in terms of a subject, but a way of transforming education itself. The report also stressed that '… while textbooks are important, teachers should be able to develop educational programmes and projects… this change will bring with it requirements of resources—in terms of time, in terms of enhancement of basic facilities, in terms space etc…'" (Note *EE in the formal system* sent on 5 March 2012 by Kartikeya Sarabhai, Director CEE, to Ministry of Environment and Forests, Government of India).

While *infusion* is the mainstay approach followed, it attracts criticism for diffusing the intent of EE in the curriculum.

> The current approach to EE has been centered around the infusion approach at least for two decades with the idea that the separate subject would overload students. With multiple and constant inputs to further Environmental Educational technologies and innovations made through great efforts by NCERT. This has still not shown adequate results. When this is assessed at the teacher and student level it has been found to be only marginally successful and patchily distributed. It has only worked where some influences that have been external to the regular curricular teaching has been focused specifically on EE. Questioning teachers across the country on infusion approach to EE has demonstrated that no subject teacher takes ownership for EE. Therefore at the student level there has been no strong focus on EE which has been taught in a completely fragmented way in isolated and unconnected project based activities in different subjects. On questioning students and parents they have the impression that this is a small unimportant part of various subjects that can be dropped and thus only a small percentage of students actively participate in such projects which are most frequently done in high profile urban schools (Personal Communication dated March 3rd, 2012, from Dr. Erach Bharucha Director, Bharati Vidyapeeth Institute of Environment Education and Research, Pune, to B. M. S. Rathore, Kartikeya Sarabhai, M. C. Mehta, Jayashree Sharma. Shared with the authors by Kartikeya Sarabhai, Director, CEE).

In the same communiqué, Bharucha stresses the need to address the problem of an infusion approach without actually 'abandoning' it but by supplement(ing) this with a well focused subject based approach alongside environmental studies as a core subject. Infusion would then feed into environmental studies from General science, social science, mathematics and chemistry.

In conclusion, the region needs to see both a qualitative and a quantitative expansion of EE and ESD through formal, informal and non-formal approaches. A focused and multi-strategy approach EE and ESD is needed that can free itself from the shackles of education based on borrowed models and solely for economic prosperity. A focus on economic prosperity means incomes contributing to personal improvement but not implicitly fostering equality, cultural sensitivity, ethics and values. There is a need to radically transform education through changing its philosophy to provide a new framework to suits the environmental, economical, social and cultural realities of the region.

1.5 EE and ESD Strategies and Initiatives in South Asia

In Chap. 2, the authors trace and establish how the paradigm of development has changed and influenced the paradigm of education (which needs further change) with the renewed understanding that environment and development interrelated and that each is essential for the other—development without environmental considerations is a disaster and untenable and that environment can never be a priority if there is no development. The other important influence on development and education they note is when human condition found an emphasis in 1992 at the UNCED in Rio and became of the foucs of the political agenda.

Now, the 17 Sustainable Development Goals (SDGs) provide an overarching framework for the post 2015 development agenda which is a maturation of the thinking related to environment, development and sustainability. The links are well understood now—sustainable development views economy as strongly embedded in society which is in turn is embedded in the environment. They establish that both the paradigm of development and education need to change in fundamental ways (Sarabhai 2005).

Joshi and Yadav, in Chap. 3 talk about sustainable development as empowerment of people and the key to empowerment process being action-oriented education (Ahmedabad Declaration made at the international conference on Education for Sustainable Future in 2005). The launch of the UN Decade of Education for Sustainable Development (DESD) in 2005, they say, found a gradual resonance in the countries of South Asia. They present some of the key actions from South Asian countries that were taken in the Decade.

In Chap. 4, Salequzzaman and Gorana, note that developmental challenges like population growth, poverty, livelihoods, climate change, fragile ecosystems, that are interrelated and work through negative feedback mechanisms have widened the narrow mandate of education from preparing people for economic prosperity to address the multiple developmental challenges the country is facing. They infer that education for sustainable development (ESD), though not carried out formally under that designation, can be seen happening through various avenues of learning, important among them being non-formal education. "Non-formal education, encompassing a wide array of activities, including alternative primary schools, youth training, literacy programmes, and professional education, can be an important complement to formal education" (EPDC 2008: 1).

In the same chapter, the authors quote Shohel and Howes (2006) who shared that non-formal education programmes run by NGOs in Bangladesh focused on reducing illiteracy, contributing to the basic education of poor children, education of girls and empowerment of women, and also in supporting government's universal primary education programme. Innovative learning methods in non-formal schools are aimed at the development of practical skills, including matters of health, sanitation and literacy, to be applied in real life situations. Many non-formal education programmes seek to integrate traditional environmental wisdom with insights and skills available from the sciences. It is developmental problems that raise awareness and stimulate action.

Two aspects can be attributed to the prevalence of the non-formal education (NFE) avenue for education—one being delivery gaps (quality, access, utility, relevance of curriculum) in the formal education systems provided by the government and private systems, and the other being strong socio-economic barriers like poverty and inequity for people to take up formal education. The fact remains that the whole country of Bangladesh has to be involved in the process of sustainable development and non-formal education provides that second best chance for Education for Sustainable Development.

Conversely, it can be said that the current models of formal education emphasize theory which draws heavily from the sciences, seldom integrating the wisdom and insights of people and disregarding cultural underpinnings. There is a kind of functionality that NFE lends to learning by taking the orientations and contexts which is an essential underpinning of Education for Sustainable Development. On the face of it there is a need to focus on non-formal education as a necessary strategy for ESD, particularly for developing countries like Bangladesh.

In Chap. 10, Dhindsa talks about the challenges for adopting ESD. He states how as a new country Pakistan focused more on security issues at the cost of social sector. The Situation Analysis Study on EE in Pakistan gives a clear picture on the status of EE in the country. It talks about the need for revising the quality and quantity of content. The study points out the need for integration of more activities and improvement in the knowledge content, introduction of case studies and environmental issues into the textbooks. The Analysis envisages that such measures would help in strengthening the National Curriculum for delivering the concepts of EE. Dhindsa, also states how mainstreaming EE and ESD could be a huge challenge, particularly when madarassas—religious schools have a separate curriculum. Very typical of the South Asia region, the NGO sector is quite active in Pakistan where due focus is given to non-formal and informal education approaches.

In Chap. 7, Pandya traces the national and international developments vis-à-vis quality education and Education for Sustainable Development and finds parallels in education taking on a transformative role both in India and globally. She traces the idea of education for sustainability to Mahatma Gandhi when he envisaged a new India in his proposal Nai Taleem (New Education also known as Basic Education). This vision of a new social order was first shared in 1937, a decade before India became independent from British rule. For Gandhi, the goals of education and society were not separate. She notes how developments internationally, particularly the Jacques Delors report *Learning: The Treasure Within*, the World Commission on Environment and Development, Chap. 36 of Agenda 21 (the outcome of the United Nations Conference on Environment and Development) actually broadened the scope of education to integrate notions of population, poverty, environmental degradation, democracy, human rights and peace, development and interdependence. From these parallel influencing forces, she states India's National Curriculum Framework (NCF) (2005), attempted to address what Naik (1974) described as the 'elusive triangle' of Indian education—equality, quantity and quality. She shares her experience of the process of developing Environmental Studies textbooks for classes III, IV, V and how infusing the intent and essence of both the NCF 2005 and ESD threw challenges and provided opportunities considering that the textbook is the sole and most important tool of instruction.

The NCF 2005, she says, demonstrates the process of infusing a 'Sustainability Perspective' (ESD as the present day's articulation of Basic Education and a self-sustaining society in a global and knowledge society) into the formal school system, in the context of the developmental and educational requirements of the country.

Values are an essential component of any discussion about sustainability, especially in an educational frame of reference says Karma Ura in Chap. 5. Worldview, influence the whole environment and explores the need to integrate Gross National Happiness (GNH) values in textbooks—values education as a means to develop a set of beliefs and attitudes as a person's character and personality unfold, so that his or her emergent beliefs will influence his or her behaviour and actions in a particular manner and direction. Initially, this process requires the articulation of a set of values and this is followed by the determination of ways of applying them in personal and collective social contexts. It is also about the exploration of those common values that will contribute to the creation of a positive future for all.

Ravindranath begins Chap. 6 by recounting the famous Chipko movement of 1974 which shows that environmental education and care are not new to this region. Centuries of frugal and careful tending of land on which communities have directly depended has occurred as a result of living in harmony with nature. Each generation has passed on its profound understanding of how to manage natural resources. Movements like this have helped the government develop a people-centred ecologically sensitive natural resource policy. She quotes Indira Gandhi (former Prime Minister of India) who observed that all forms of life are closely linked and disturbance in one caused imbalances in the other. She emphasizes the mounting pressure on natural resources and unprecedented environmental degradation across the country that actually made a strong the case for EE to redefine development for a sustainable future. She mentions how the highest court of the country, the Supreme Court, accepted in principle that "through the medium of education awareness of the environment and its problems related to pollution should be taught as a compulsory subject" in 1991 (Sarabhai 2004).

In Chaps. 6 and 7, Ravindranath and Pandya respectively draw the pattern of the growth graph of EE in India which they trace alongside international thinking. They both observe how the scope of EE has broadened in India along with educational and societal developments.

Ravindranath showcases many interesting initiatives of the Ministries of Environment and Forests and Human Resource Development of Government of India. She charts how the change from traditional-knowledge based environmental movements, to constitution based rights and duties and to the judicial route to inculcating environmental awareness through education. This has culminated, she notes, in the cutting edge initiatives of the Ministries of Environment and Forest and Ministry of Human Resource Development, and last but not least, the role of media, corporations, NGOS, and communities which have collectively contributed to the changing paradigm of EE in the country. She recommends that in India the scope of ESD has to be clearly defined and stated in all the state policies, programmes and project plans so that ESD is not brushed aside like EE as a supplementary school subject.

In Chap. 9, Pande discusses at length the process of integration of EE into the school level curriculum in Nepal. The initiatives at the grassroots level, with support from various donor agencies and ministries of the government, have helped in influencing educational policy makers, planners, and other stakeholders. He shows how international non-governmental organizations (INGOs) have played a pivotal role in the promotion of EE in Nepal. He describes how education has been reoriented for poverty alleviation, eradication of illiteracy, improving the access of disadvantaged groups to education, developing human resources in accordance with national development requirements, and making education contextual and relevant to life situations thus establishing the link between education and development.

The experience in Nepal and other countries of South Asia with regard to EE and ESD initiatives brings to the fore the observation made by Chi-Kin Lee and Williams that "International declarations may be praised for their flexibility in that they allow for adaptation to local circumstances. They may, at the same time, be criticized for their lack of precision and over-ambitious unreality" (2009: 26). Governments and policy-makers piggy back on existing systems, and this necessitates inputs in terms of enabling policies, capacities, and budgets. "In international terms, proposals for concrete action need to be supported by the resources necessary for achieving any specific targets" (2009: 26).

Social cohesion is a way for peace notes Sedere in Chap. 11 on Sri Lanka. Educational interventions initiated in Sri Lanka to close the gap and bring better social cohesion through education are highlighted. Increasing evidence shows that social cohesion is critical for societies to prosper economically and for development to be sustainable (http://web.worldbank.org/WBSITE/EXTERNAL/TOPICS/EXTSOCIALDEVELOPMENT/EXTTSOCIALCAPITAL/0,,print:Y~isCURL:Y~contentMDK:20185164~menuPK:418217~pagePK:148956~piPK:216618~theSit ePK:401015,00.html. Accessed on December 11, 2015) Smith in his discussion on social capital elaborates its significance for educators. "Child development is powerfully shaped by social capital. Trust, network, and norms of reciprocity within a child's family, school, peer group, and larger community have far reaching effects on their opportunities and choices, educational achievement, and hence on their behaviour and development" (http://infed.org/mobi/social-capital/. Accessed on May 6, 2013). By bringing these educational initiatives to the school, schools and students can become the entry point for this useful organizing idea. The World Bank (1999) has also brought together a range of statistics to make the case for the social and economic benefits of social capital. For example, they argue that there is evidence that schools are more effective when parents and local citizens are actively involved. "Teachers are more committed, students achieve higher test scores, and better use if made of school facilities in those communities where parents and citizens take an active interest in children's well being."

Sedere writes about social contracts and conflict resolution education to cultivate and regain ethnic tolerance for one's fellow citizens, while developing opinions and attitudes of the children, families and communities to the ethnic diversity of the country. Sri Lanka is also focusing on social contracts to wipe out the mistrust and fear of each other by the ethnic groups produced a generation of youth who regarded each other as cultural 'strangers (Smith 2000–2009) also records that a growing body of research suggests that where trust and social networks flourish, individuals, firms, neighbourhoods, and even nations prosper economically. Social capital can help mitigate the insidious effects of socio-economic disadvantage.

Shareef, in Chap. 8, shows how the Ministry of Education in the Maldives and UNICEF have been working to transform the education system from teacher-centered rote learning to child-centered active learning so that children become more actively engaged in their education. A deep rooted phenomenon observable in the South Asian region is that schools and learners strive hard throughout the academic calendar to prepare for examinations—good results in the examinations are considered as an achievement. This unique phenomenon/trend is very influential for various factors related to education and learning in the schools. Curricula are book-based, with little room for honing skills and competencies; they discourage vertical and horizontal learning across subjects and grades in the already tightly arranged subjects resulting in glaring gaps, duplications in curricula, not requiring teacher interaction proving antithetical to basic tenet of quality education like holistic learning. Teachers become transmitters of knowledge turning teachers into content experts leaving little room to believe that learning has to lifelong. Teachers are seen as instrumental in learning success of their students. These practices which have become norms of education praxis demand serious reconsideration/questioning of transmitting certainties.

The formal education system across the region requires nuanced changes—with stress on the process of learning rather than merely on knowledge as the end product. ESD, acknowledges that "all partners in the learning enterprise offer students the opportunity to actively investigate objects, events, text materials, people, places, and phenomena" (Audet and Jordan 2008: x).

Advocates of inquiry-based teaching and learning, as taken up in the Maldives, according to Shareef, have to pay special attention to motivational factors, provide opportunities for social interaction, and create active learning environments. In such settings, traditional classroom roles for students and teachers are blurred, and formats for assessment are multiple, varied, and carefully aligned with the relevant content standards. As ultimately it allows for different assessment format, it loosens the grip of examination and assessment systems based on rote learning. Such a shift allows the acquittal from the (examination)-tail-wagging-the-dog (learning) syndrome prevalent in the region. It is in this situation that the role of the teacher gets redefined from knowledge dispenser to co-creator of the knowledge. Audet and Jordan (2005) argues that "The principle factor that determines the level of inquiry is the relative amounts of student versus teacher control over an activity." Further, he assures that, "the skills, processes, tools, and elements of inquiry are developmental. Skillful teachers know that choosing the most appropriate instructional strategy is influenced by the time of year, age level, amount of experience, and nature of the learning activity."

In conclusion, the Chapters in the book, show how ESD is taking root and shape in the South Asian countries. The contexts have influenced the meaning and agenda for ESD. The context is critical in the reorientation process of education and learning.

References

Abe, O., & Bhandari, B. (2000). *Environmental education in the Asia-Pacific Region: Some problems and prospects*. Kanagawa: IGES.

Abe, O., & Bhandari, B. (2001). *Environmental education in the Asia Pacific Region: Status, issues and practices*. Kanagawa: IGES.

Audet, R., & Jordan, L. (2005). *Integrating inquiry across the curriculum*. Thousand Oaks: Corwin Press.

Audet, R. H., & Jordan, L. K. (Eds.). (2008). *Integrating inquiry across the curriculum*. Heatherton, Vic: Hawker Brownlow Education.

Bandhu, D., Singh, H., & Maitra, A. (1989). Conference: *Environmental education and sustainable development*. New Delhi: Indian Environment Society. https://environmentalartilces.wordpress.com/indira-gandhiji-1917-%E2%80%93-1984-environmentalist/. Accessed 8 Dec 2015.

Banerjee, S. B. (1999). *Sustainable development and the reinvention of nature*. Paper presented Critical Management Studies Conference (Environment Stream) Manchester, UK. www.mngt.waikato.ac.nz/ejrot/cmsconference/1999/…/environment.pdf

Chaudhary, L. (2007). *An economic history of education in colonial India*. Stanford: Hoover Institution, Stanford University.

Education Policy and Data Centre. (2008). *The extent and impact of non-formal education in 28 developing countries*. Washington, DC: EPDC.

EFA. (2015). Ministry of Primary and Mass Education, Government of Bangladesh, Education for All 2015 National Review Report: Bangladesh. Submitted to UNESCO. http://unesdoc.unesco.org/images/0023/002305/230507e.pdf. Accessed on 8 Dec 2015.

Fergus, A., & Rowney, J. (2005). Sustainable development: Lost meaning and opportunity? *Journal of Business Ethics, 60*, 17–27.

Gutman, L., & Akerman, R. (2008). *Determinants of aspirations*. London: Centre for Research on the Wider Benefits of Learning Institute of Education.
Hesselink, F., Van Kempen, P., & Wals, A. (2000). *ESDebate international debate on education for sustainable development*. Gland/Cambridge: IUCN.
Hopkins, C. (2011). ESD: One of the many contributions in creating a more sustainable future. *Asia-Pacific forum on educational cooperation: Synergies and linkages of EFA, ESD and ASPnet for sustainable Asia and the Pacific (Tokyo, Feb 19–21, 2011)*. Japan: UNESCO. http://hdr.undp.org/en/content/human-development-report-2011. Accessed 6 May 2013 http://www.accu.or.jp/esd/forum_esd_2010/program/program12_01/pdf/text2.pdf. Accessed 7 May 2013.
Human Development Report. (2011). *Sustainability and equity: A better future for all*. UNDP. http://www.unescobkk.org/education/news/article/reforming-international-cooperation-in-education/. Accessed 7 May 2013.
Institute for Global Environment Strategies. (1999). *International conference on environmental education in the Asia Pacific Region*. Kanagawa: IGES.
Kazuo.(2011).http://www.unescobkk.org/education/news/article/reforming-international-cooperation-in-education/. Accessed on 7 May 2013.
Lee, L., & Williams, M. (2009). *Schooling for sustainable development in Chinese communities: Experience with younger children*. Volume 1 of Schooling for Sustainable Development. USA: Springer Netherlands.
Naik, J. P. (1974). *The role of government of India in education*. New Delhi: Government of India.
National Council of Educational Research and Training. (2006). *National focus group on habitat and learning*. New Delhi: NCERT. (Position Paper).
Nepal National Education Planning Commission. (1956). *Education in Nepal: Report of the Nepal national education planning commission*. Kathmandu: Bureau of Publications, College of Education.
Nikolopoulou, A., Abraham, T., Mirbagheri, F. (2010). *Education for sustainable development: Challenges, strategies and practices in a globalizing world*. New Delhi: Sage Publications. NCERT. http://www.ncert.nic.in/html/pdf/educationalsurvey/Manual_on_Statistics_and_Indicators_of_School_Education/Annexure___IV___Glossary_of_Importnat_Terms.pdf. Accessed 8 Dec 2015.
Palmer, J. (2002). *Environmental education in the 21st century: Theory, practice, progress and promise*. London: RoutledgeFalmer.
Ross, R., & Savada, A. (1988). *A country study: Sri Lanka*. http://countrystudies.us/sri-lanka/46.htm. Accessed 7 May 2013.
Roy and Sharma (1991). *English for students of science*. New Delhi: Creative Advertisers.
Saeed, S., Goldstein, W., & Shrestha, R. (1998). *Planning environmental communication and education: Lessons from Asia*. Bangkok: IUCN.
Sarabhai, K. V. (2000). *Securing our future in the new century: Lessons from India*. New Delhi: IUCN.
Sarabhai, K. V. (2004). *Thoughts on environmental education at the school level*. New Delhi: NCERT. http://www.greenteacher.org/images/EE%20Papers/Thoughts%20on%20Environmental%20Education%20at%20the%20School%20Level.pdf. Accessed 3 May 2013.
Sarabhai, K. V. (2005). *Education for sustainable future*. http://www.ceeindia.org/esf/esf.asp Accessed 15 Dec 2015.
Sarabhai, K. V. (2009). *India: A Commitment to ESD*. Towards a New Development Paradigm, Education for Sustainable Development, India Report to the World Conference on ESD, Bonn, Germany, March 31 to April 2009, Indian National Commission for Cooperation with UNESCO, Ministry of Human Resource Development, Government of India, CEE.
Sarabhai, K. V. (2010). ESD in a developing nation. *Tomorrow Today*. Paris: UNESCO (Published by Tudor Rose on behalf of UNESCO).
Savalia, R. (2011). *NFE as an ESD: Initiatives and perspectives*. Synergies and Linkages EFA, ESD and ASPnet, Final report of Asia-Pacific forum on Educational Cooperation. Tokyo, Japan: ACCU

Shohel, M. M. C., & Howes, A. J. (2006). Non-formal education for sustainable development: A Bangladeshi perspective. In: *The 10th APED international conference: Learning together for tomorrow, education for sustainable development*, Bangkok, Thailand. http://www.unescobkk.org/education/apeid/apeid-international-conference/10-th-apeidinternational-conference/. Accessed on 29 Mar 2012.

Smith, M. (2000–2009). 'Social capital', *the encyclopedia of informal education.* http://infed.org/mobi/social-capital/. Accessed 6 May 2013.

South Asia and Southeast Asia Network for Environmental Education (SASEANEE), CEE. http://www.saseanee.org/. Accessed 6 May 2013.

South Asia Co-operative Educational Programme. *Mission.* http://www.sacep.org/html/about_visionmission.htm. Accessed 6 May 2013.

Thaman, K. (2002). Shifting sights: The cultural challenge of sustainability. *International Journal of Sustainability in Higher Education, 3*(3), 233–243.

The World Bank. (1999). *Social capital.* http://web.worldbank.org/WBSITE/EXTERNAL/TOPICS/EXTSOCIALDEVELOPMENT/EXTTSOCIALCAPITAL/0,,contentMDK:20185164~menuPK:418217~pagePK:148956~piPK:216618~theSitePK:401015,00.html. Accessed 29 Apr 2014.

Tuladhar, G. *Status of education in Nepal and development plan.* http://web.isc.ehime-u.ac.jp/ice/6-1@Gangalal%20Tuladhar_text%20_8p.pdf. Accessed 7 May 2013.

United Nations Educational, Scientific and Cultural Organization. (1997). Conference: *International conference environment and society: Education and public awareness for sustainability.* portal.unesco.org/…/d400258bf583e49cd49ab70d6e7992f6Thessaloniki. Accessed 6 May 2013.

United Nations Educational, Scientific and Cultural Organization. (2005). *A situational analysis of education for sustainable development in the Asia-Pacific Region*. Bangkok: UNESCO.

United Nations Educational, Scientific and Cultural Organization. (2006). *Highlights on progress to date January 2006*. Paris: UNESCO.

United Nations Educational, Scientific and Cultural Organization and United Nations Environmental Programme. (1989). Developing an environmental education curriculum. *UNESCO-UNEP Environmental Education Newsletter*. Vol. XIV, No. 3, September. Paris: UNESCO.

United Nations Educational, Scientific and Cultural Organization and United Nations Environmental Programme. (2007). *The Ahmedabad declaration 2007: A call to action – Education for life: Life through Education*. Ahmedabad: UNESCO with UNEP. http://aries.mq.edu.au/pdf/AhmedabadDeclaration.pdf. Accessed 24 Nov 2015.

World Commission on Environment and Development. (1987). *Our common future*. Oxford/New York/New Delhi: Oxford University Press.

Chapter 2
A New Paradigm of Education Towards Sustainable Development

Rajeswari Namagiri Gorana and Preeti Rawat Kanaujia

2.1 Sustainable Development

Respect for environment and sustainability are not particularly new and in fact are well enshrined in the collective values, social institutions, religious beliefs and practices, social enterprise of many ancient cultures, including those from the South Asian region. Environment as the foundation of all life and lifestyles is reflected in the respect for environment. Ecological integrity or the conception of sustainability—a word well recognized and understood word today—could be found in this the hymn c.1200 B.C. Atharva Veda (an ancient Indian scripture) hymn 'What of thee i dig out, Let that quickly grow over, Let me not hit thy vitals, Or thy heart'. The guidelines for sustainability were established long before it emerged as a concern of modern living. 'The earth has enough for everyone's need but not for anyone's greed', said Mahatma Gandhi. Sarabhai and Chhokar (2009; p. 51) also observe how Gandhi spoke of the need to conserve the unique life support systems long before the recognition of the environmental crisis. (Taylor et al. 2009: 51)

Internationally, the idea of sustainable development goes back at least to the 1970s. In 1972, the United Nations Conference on the Human Environment held in Stockholm brought the industrialized and developing nations together to delineate the 'rights' of the human family to a healthy and productive environment (http://www.uncsd2012.org/history.html#sthash.1TQMWb.dpuf. Accessed on December 10, 2015). The precursor to the concept could be the inherent tensions between eco-

R.N. Gorana (✉)
Programme Coordinator, Children's Media Unit, Centre for Environment Education (CEE), Ahmedabad, India
e-mail: rajeswari.namagiri@ceeindia.org

P.R. Kanaujia
Senior Programme Coordinator, Northern Regional Cell, Centre for Environment Education (CEE), Lucknow, India
e-mail: preeti.rawat@ceeindia.org

R.N. Gorana, P.R. Kanaujia (eds.), *Reorienting Educational Efforts for Sustainable Development*, Schooling for Sustainable Development,
DOI 10.1007/978-94-017-7622-6_2

nomic development and environmental protection when each was seen as independent entity or having the potential to undermine or interfere with each other. The conception of sustainable development delineated the two-way connection between economic development and environmental quality—one being that development that ignores environment will be a disaster and the other as put forth by Indira Gandhi "…the environment cannot be improved in conditions of poverty" (Gandhi 1972: 36). The reference to poverty was articulated once again in the World Conservation Strategy (WCS) brought out by International Union for the Conservation of Natural Resources (IUCN) in 1980. "The Strategy asserted that conservation of nature cannot be achieved without development to alleviate poverty and misery of hundreds of millions of people and stressed the interdependence of conservation and development in which development depends on caring for the Earth"(http://www.uncsd2012.org/history.html#sthash.1TQxMMWb.dpuf Accessed on December 10, 2015).

Human condition found an emphasis and became of the focus of the political sustainable development agenda. The sustainable development movement that came out of the late 1980s sought action to improve the living standards of those who are in need. "Many of the advanced countries of today" Gandhi said, "have reached their present affluence by their domination over other races and countries, the exploitation of their own masses and their own resources…Now as we struggle to create a better life for our people, it is in vastly differing circumstances, for obviously in today's eagle eyed watchfulness, we cannot indulge in such practices even for a worthwhile purpose". (Gandhi 1972: 36)

The Director of the United Nations Environment Programme (UNEP), Klaus Töpfer, noted in 2001 that:

> In her foreword to Our Common Future in 1987, Gro Harlem Brundtland commented that the period since Stockholm was a time of "standstill or even deterioration in global cooperation," and "a retreat from social concerns." Notwithstanding growing scientific evidence of environmental crises no-one was prepared to act (UNEP 2001: 6).

One might not go as far as Brundtland's condemnation of progress in the 1970s and early '80s but certainly there seemed no more than a slow maturation of ideas. Sustainable development, as a concept that espoused a fine balance between needs and wants—including that of the present and future generations took hold through World Commission on Environment and Development (WCED)'s report, Our Common Future (Gro Harlem Brundtland was the chairman of the World Commission on Environment and Development). The goal was the better management and protection of the ecosystem and the creation of a more prosperous future for all. It highlights the need to improve the quality of people's lives in a manner that doesn't diminish or compromise the environment's capacity to provide for the future.

The concept of sustainable development carried with it the premise that economic, environmental and social well being are interrelated and changed the discourse of development cannot limit alone to economic variables and that it needs to include socio-cultural, political and environmental variables. As is noted later this became a catalyst for action.

The United Nations Conference on Environment and Development (UNCED) popularly known as the Earth Summit, held in Rio de Janeiro in 1992, brought sustainable development to the international centre stage with a clear message that transforming attitudes and behaviour were prerequisites to bring about changes necessary to ensure a sustainable future for all people. Environment and development are perceived as integral parts of sustainable development where economic development is considered as a contributing factor for human development but not an end in itself. However, varying strengths of emphasis on development than environment has led to different ways of interpreting and applying this concept to policy formulation.

It would be appropriate to mention India's report to the UNCED prepared by Centre for Environment Education—the organization the editors of this book come from—notes that "the real challenge for a country like India was not 'how to get there' but 'how not to'. By this we meant that while the dominant development paradigm was pulling all developing nations towards achieving living standards that emulated the largely unsustainable lifestyles of the Western developed nations, the challenge for these nations really was to break away from a development approach that was merely imitative of the West and to embark on an alternative development pathway by making choices that were indeed different, appropriate and sustainable to 'pick right'." (Sarabhai 2010: 69)

The journey from Stockholm to Rio, and to Johannesburg and beyond has taken more than 30 years and seems to find a firm footing in the Sustainable Development Goals (SDGs) with a global acceptance of the 17 goals as a framework for development. Markedly, the discourse on sustainable development has reached a comprehensive level with the launch of Sustainable Development Goals (SDGs) as part of the post-2015 development agenda. It was during the UN Conference on Sustainable Development in 2012, the governments agreed to launch a process to develop a set of SDGs. The document—formally adopted at a UN summit of world leaders on September 25–27—2015 includes the 17 Sustainable Development Goals, supported by 169 specific targets.

Sustainable Development Goals

Goal 1	End poverty in all its forms everywhere
Goal 2	End hunger, achieve food security and improved nutrition and promote sustainable agriculture
Goal 3	Ensure healthy lives and promote well-being for all at all ages
Goal 4	Ensure inclusive and equitable quality education and promote lifelong learning opportunities for all
Goal 5	Achieve gender equality and empower all women and girls
Goal 6	Ensure availability and sustainable management of water and sanitation for all
Goal 7	Ensure access to affordable, reliable, sustainable and modern energy for all
Goal 8	Promote sustained, inclusive and sustainable economic growth, full and productive employment and decent work for all

Goal 9	Build resilient infrastructure, promote inclusive and sustainable industrialization and foster innovation
Goal 10	Reduce inequality within and among countries
Goal 11	Make cities and human settlements inclusive, safe, resilient and sustainable
Goal 12	Ensure sustainable consumption and production patterns
Goal 13	Take urgent action to combat climate change and its impacts
Goal 14	Conserve and sustainably use the oceans, seas and marine resources for sustainable development
Goal 15	Protect, restore and promote sustainable use of terrestrial ecosystems, sustainably manage forests, combat desertification, and halt and reverse land degradation and halt biodiversity loss
Goal 16	Promote peaceful and inclusive societies for sustainable development, provide access to justice for all and build effective, accountable and inclusive institutions at all levels
Goal 17	Strengthen the means of implementation and revitalize the Global Partnership for Sustainable Development

(http://www.un.org/ga/search/view_doc.asp?symbol=A/RES/70/1&Lang=E. Accessed on December 10, 2015).

As, we have seen here, sustainable development views economy as strongly embedded in society which is in turn is embedded in the environment. To achieve sustainable development goals, there need to be among others policy changes, changes in the systems of work, and changes in the technologies we use. This involves not only a change in understanding but, in addition, changes in the perspectives and values used in learning about development.

> …policy instruments or technological solutions are not going to be enough and ….behavioural change (is) critical to achieving sustainable development. Thus, the role of education in its broadest sense including training and capacity building, communication and creating public awareness becomes a key strategy for achieving the Sustainable Development Goals (Sarabhai 2015, p. 123).

2.2 Education Towards Sustainable Development

"Education for Sustainable Development is born out of a very simple idea: reaching sustainability will require more than legal frameworks, financial resources and green technologies; it also needs us to change the way we think--change that best can be obtained through education" (UNESCO 2010: 4).

The power of education as a driver of Sustainable Development was realized when an explicit link between education and sustainable development was made in Rio Summit in 1992. Chapter 36 enshrined in Agenda 21 promoted education and public awareness of environment and development. Countries, signed up to work for the Agenda 21 commitments. It further served as a calling for a reorientation of education towards sustainable development.

The 1997 international conference in Thessaloniki on 'Environment and Society: education and public awareness for sustainability', third in the series of intergovernmental Environment Education conferences organised by UNESCO (1977: The Intergovernmental Conference on Environmental Education, Tbilisi; 1987 International Congress on Environmental Education and Training, Moscow; (Sarabhai et al. 2007), emphasised the transition from environmental education (EE) to education for sustainable development (ESD).

For those working in the field, ESD was just a refocusing of environmental education but for others there was a new emphasis bringing together development education and environmental education by adding social and economic concerns (Hogan and Tormey 2008). Consequently ESD varies greatly in practice with some practitioners concerning themselves with ecology at its heart whereas others take a more balanced perspective across the interplay of ecology, society and economy (Lee and Efird 2014) or even focus on education for development *per se* (McKensie et al. 2015).

It goes on to note that Chapter 36 of Agenda 21 identified four major thrusts for ESD: (1) improve access to quality basic education, (2) reorient existing education to address sustainable development, (3) develop public understanding and awareness, and (4) provide training programmes.

"Unfortunately, in many countries, the current level of basic education is low and the quality so lacking that it severely hinders national options and plans for a sustainable future" (p. 5). For this reason there is frequently emphasis on basic education but this is sometimes at the expense of the other three and the implications of quality education are forgotten. UNICEF's Anandshala (Schools of Joy) improved the infrastructure of a school and the quality of teaching-learning process. Activities include school infrastructure improvement, capacity building of teachers and parents towards quality education and participatory approaches, development of school master plans to make schools joyful places. The Anandshala schools have capacity-built the teachers, parents and students in becoming environmental stewards. The Anandshala model is a testimony that education can build capacities to develop environmentally conscious students. Basic education is more than just improving the workforce for today, but is a means to an end which addresses environmental stewardship for the longer term.

Sarabhai (2005) notes that "not only the paradigm of development that needs to be changed to achieve sustainability, but that the paradigm of education also has to change in certain fundamental ways" and that this is not widely recognized (http://

www.ceeindia.org/esf/esf.asp Accessed on December 15, 2015). "The various points are interrelated and form part of a single paradigm shift which is needed as we move towards education for sustainable development.

Learning Rather Than Teaching
This change is fundamental to the way one thinks of education—Education, from being something that society determines for its citizens, or adults decide for their children, will have to become "seeker" oriented. The classical "guru" concept of ancient India where the guru would teach, as strictly as a teacher, but recognizing that the pupils would someday seek their own path. Teaching, therefore, needs to be focused on making the student a better learner, rather than filling the student with information per se. And ESD is not only about learning individuals, but learning institutions and societies.

Lifelong and Continuous Rather Than Confined to a Specified Period
Education was something thought of as an activity you essentially did during the first quarter or so of your life. It aimed to give you the knowledge and skills required for the rest of your life. With the rapid pace of change today, this is obviously not quite enough. Institutions have started a variety of in-service courses. Increasing adult education programmes are available. Non-formal opportunities and opportunities for community education have increased manifold. But the older paradigm still remains largely intact. In a new world through internet and other emerging educational opportunities, it is likely that certification will no longer be based on "residency" factors. Technology is also driving people to learn new skills. But there are still many who have a diffident attitude towards new learning. "I'm too old" to learn. Distance learning is emerging as a major alternative way for learning. But issues such as access for all and digital divide are continuing concerns.

Multi-sourced and Accessed Rather Than Top Down, Controlled, and Orchestrated
The school and the textbook no longer have the virtual monopoly they had on the child's mind. Most children have independent access to information, without 'gate keepers'. Even parents can't 'control' the educational exposure the child has. And this is only going to increase. The teacher now has to 'compete' with other sources. The student can verify information given in the class room.

Empowering Rather Than Socializing (Indoctrinating)
ESD is about individuals and communities, questioning, visioning and making changes. It is, in that way, contrary to conventional education which is a process of a system meant for fitting people to fit in and not upset the status quo, to a process where people are empowered to change the world.

Global and Yet Locale Specific
Education today needs to give people the global perspectives and connections, at the same time enable them to interpret generic learnings in their local context, and to not only act locally, but see the wider impacts of their actions.

Capacity Building to Build Abilities for Critical Thinking and Problem Solving
With the onslaught of information from various sources and ever increasing amounts of data, the issue is often of being able to select and process information, critically analyze it, and take decisions based on this. It is these skills that ESD attempts to develop.

Multi-disciplinary Approach as Opposed to a Single New Discipline
Real world problems need integrated, multi-disciplinary solutions while most institutions—from school education on, and including government—are organized sectorally. This compartmentalization starts with school education. An important task for ESD is to break these barriers, and help in seeing connections and the holistic linkages.

> **Sensitive to Gender, Diversity etc.**
> ESD needs to not only be sensitive to, but also support and enhance the celebration of diversity—of cultures, languages, societies, and diversity in the natural world. Sensitivity to gender, equity and other issues, needs to be an integral part of ESD, and it needs to be inclusive of, as against discriminatory on, gender, caste, and other factors.
>
> **Participatory and Based on Learning with Peers**
> Conventional education dictates what and how learners should learn. In an increasingly complex world, people will decide what they need to learn and how best, and when, they need to learn. ESD needs to empower and enable learners to learn when, where, what and how they choose, as opposed to one-directional teacher to passive learner."

While government commitment to change and action is necessary for sustainable development it is clear that for any initiative to work, civil society in general and individuals in particular need to be involved at all stages but at times there seems a reluctance for government and educational bodies outside the ESD lobby to become engaged. There is a need to drive, "…ESD out of its circle of believers into other fields and disciplines to carry out the mandate of change for sustainability" (Nambiar 2014, p. 88).

The Bonn Declaration (UNESCO 2009), nonetheless, emphasised the interplay environment, society and economy and the role of education in "securing sustainable life chances, aspirations and futures for young people" (p. 2, point 5). Sustainability in the context of development is an intensely agenda-laden concept and as such Education that promotes sustainable development is squarely aimed at change (Nambiar and Sarabhai 2015). Two paradigms that have been proposed are the moralistic and the educational (Schusler et al. 2009). The first seeks to modify the learners' lifestyle behaviours in a predetermined direction in order to directly contribute to solving problems and generate change.

The educational paradigm, on the other hand, is more open ended and less direct, stressing the importance of strengthening the capabilities of learners to reflect and take a standpoint that will similarly bring about change (Almers 2013). These discourses of change within ESD parallel the educational debate about transformative education which is required if current global challenges are to be addressed (Gadotti 2010).

This sees an education that focuses on critical reflection, holistic approaches, relationships and futures. It is concerned more with why we are teaching than with how or what we teach; ethical development is at the core (Pavlova 2013). Noguchi et al. (2015) helpfully use the metaphor of the muscat grape to emphasise the key action principles and supporting mechanisms for ESD.

> Each cluster (of berries) is made up of grape berries that represent one of the six action principles, namely: responding, engaging, enabling, embedding, sustaining and transforming ourselves. The four policy support mechanisms are: providing resources, networking/partnership, capacity building and content development (p. 9).

This seems deceptively simple which it is not. "Sustainable development is context-specific and needs educational inputs at every stage, starting with recognizing the need for sustainability, negotiating what sustainability means to the individual and the community, seeking technical information and training to be able to identify and address sustainability issues" (Nambiar and Sarabhai 2015, p. 3).

New, unfamiliar situations which underpin sustainability but are difficult to envision and which do not immediately relate to current, familiar, physical experiences are understood by individuals through their metaphorical, conceptual systems. Thus to be successful ESD needs to create cognitive bridges to learners' prior experiences and to happenings or people to which they are attached and make reference analogies that allow people to imagine what might happen in the future" (Dickson et al. 2013, p. 146).

Knowledge of the science does not necessarily translate into recognition that the environment is at risk or the propensity by individuals to take action. Individuals may agree with the scientific evidence presented but resist changes in their personal behaviour which is essential. Talk of the absence of sustainability may likely threaten people's current and future lifestyle, health, and survival, but framing the problem within such a context of threats, a strategy known as "fear appeals" (Dickson et al. 2013, p. 148) can have the opposite of the intended effect. Framing which is positive and focuses on the future of generations to come encourage individuals to see their lives as meaningful and educational opportunities worthwhile.

ESD has become interlinked with the movement towards 'Quality Education'. UNESCO (2004) describes it in the following terms, namely that:

> Quality education is an effective means to fight poverty, build democracies, and foster peaceful societies. Quality education empowers individuals, gives them voice, unlocks their potential, opens pathways to self actualization, and broadens perspectives to open minds to a pluralist world (p. 2).

ESD one can argue is not about educating people about the nature of sustainable development per se but more broadly aims to provide the basis to enable people in different settings to address the challenges that threaten the economic, social, environmental and cultural realms of their society at local and global levels. Sustainability, as Willow (2015) argues for environmentalism, needs to be seen as a way of positioning ourselves in relation to the world around us. It is about cultivating a philosophy of action which determines an individual's approach to environmental decision making (Goralnik and Nelson 2011). It means not just understanding the nature of unsustainable paths within life and the structures and systems that support them, but re-assessing and re-thinking these pathways under a new paradigm to leverage different kinds of questions that lead to new solutions.

The aim is that young people seek to develop a lifestyle where actions are consciously and reflectively driven by values and a vision for the world. It is necessary for the young person to ask, ''How can my actions reflect who I am, what I believe, and what I value?" (Goralnik and Nelson 2011, p. 182) Through this kind of efforts, they go onto contend, depends on students' knowing, "that right action fits into a wider philosophy of action that is greater than the sum of individual acts, is more meaningful than simple and finite behavior shifts." The initiative 'Young Leaders for Change' supported by Ministry of Environment and Forests, Government of India was designed to capacity build chosen young people in schools to ask the right questions to foster change not only at the individual level but at the level of the community. They worked around an environmental concern and demonstrated that 'action' can follow 'knowing' through action projects. In other words successful actions towards a more sustainable world will not be encouraged within education unless the analytical framework and underpinning values fit the broader base of the young person. This is not to say that a person's philosophy in respect a particular aspect of life including the sustainability is necessarily the same; it may or may not be but in the mind of the individual and the values that shape judgment there is compatibility.

As we have argued, ethics are central although this is not unproblematic as for example Sund and Ohman (2014) note there is the issue "whether or not the state should promote certain 'desirable' values through compulsory education" (p. 639). Notwithstanding this dilemma, a paradigm shift in education is needed in order to emphasize critical thinking, problem-solving, holistic vision, systems thinking and futures-oriented thinking. Participation, collaboration and argumentation are at the core. The logic of interconnectedness extends as well to interdisciplinary approaches (Lawale and Bory-Adams 2010). The aim is to encourage intellectual curiosity alongside the desire and capacity to act for the environment. Action competence is a useful concept (Mogensen and Schnack 2010). Philosophically,

> …the main point of action competence is the idea of action. Inspired by analytic philosophy concerning explanation and understanding and philosophical psychology as well as pragmatist analyses and critical theory the point can be made that human action differs from, or is a special kind of, mere behaviour and activity. Not only are actions intentional, the intentions, motives and reasons all have an intrinsic relation to the actions. So it will be a different action if the intention turns out to be different (p. 59).

It is "a capability—based on critical thinking and incomplete knowledge—to involve yourself as a person with other persons in responsible actions and counteractions for a more humane world" (Schnack 1994, p. 190). Action competence is a complex and dynamic educational ideal, and, according to Schnack (2000, p. 107), not possible to operationalize "by converting it into a set of observable phenomena." It puts focus on environmental ethics and is in line with the view of ESD dominated by pluralistic, emancipatory or transactional forms of education that encourage co-creation of knowledge and multiple perspectives (Kopnina 2012). A case in point is the Handprint. Action competence, as a value, is what the Handprint espouses. Sustainability is not a prescription and it that sense the Handprint symbol reminds of the prerogative to adopt lifestyle and behaviours in consonance with sustainability in any context (The Handprint was launched in 2007 during the fourth International Conference on Environment Education at Ahmedabad and symbolizes action for sustainability). Open endedness is central. ESD becomes not an end in itself but a channel for the learner to make sense of human life, what it means to live in a world that can sustain itself in the future and how problems in the environment can be overcome for the good of all. The aspiration is for young people to believe nothing unless it agrees with their reasoning, their beliefs of how to live a good life and the evidence they see.

ESD is education that understands the past, is relevant to the present, and has a view to the future (Pigozzi 2010). In thinking about this trajectory for learning in ESD the Delors (1996) four pillars: t*o know, to do, to live together* and *to be* provides a useful avenue for thought and was taken up by United Nations Economic Commission for Europe Steering Committee on Education for Sustainable Development in Geneva (Economic Commission for Europe 2011). The first pillar, *Learning to know*, implies learning how to learn about the interplay of environment and development by developing powers of reasoning, and the ability to think in a critical way in order to find solutions to complex problems. It is less concerned with the ' *acquisition of structured knowledge*' but rather the acquisition of tools to handle knowledge throughout life. The aim is to help learners understand through engagement in real-world issue that there may be a need to transform current lifestyles and that young people can make a difference. *Learning to do* involves dealing with decision-making, uncertainty and risk-taking. Critical thinking and values' clarification and the capacity to solve real problems are central with the aim developing active citizenship in participatory and learner-centred education. Interdependence puts focus on *Learning to live together* to resolve conflicting situations and respecting the value systems of others. Educational experiences will challenge unsustainable practices across educational systems in order to help learners clarify their own and others worldviews and recognize alternative frameworks. Finally there is a need to promote a person's ability to develop aspirations, goals, commitments and values in *Learning to be.*

Vision for the sort of world we want for those who will follow is critical and the ethical dimension is at the forefront of endeavours towards ESD. It is perhaps concerned more with why we are teaching (and the purpose of learning) than with how or what we teach (Mahruf and Shohel 2011) and makes no distinction between formal and non-formal action learning taking whatever opportunity arises. It requires a teacher who is a critically reflective practitioner willing to challenge assumptions underlying unsustainable practice and act as a facilitator and participant in the learning process to inspire creativity and innovation.

The heart of ESD is engaging learners as individuals with sustainability and empowering them for change by asking them to consider issues of a profoundly personal nature in relation to their own values and likely actions (Murraya et al. 2014). Almers (2013) stresses the importance of developing self-esteem, self-efficacy, participation/belongingness, authenticity, and perceived competence as central motivational factors if change is to take place. The Paryavaran Mitra programme—network of 20 million children which was implemented by CEE supported by Ministry of Environment and Forests, Government of India and ArcelorMittal was aimed to capacity build students to demonstrate environmental citizenship qualities through positive change in behaviour and action by working in five thematic areas: Water and Sanitation, Energy, Biodiversity and Greening, Waste Management, and Culture and Heritage. There emerged many young leaders who were involved in from the Paryavaran Mitra programme (Pandya and Gorana 2011). The key is the young person asking questions of him or herself and of the worlds of others understanding albeit provisionally how lives are and should be organised in searching for pathways towards a sustainable world. This personalisation of learning requires a move away from didactic pedagogies towards more open ended, values laden, learner-centred approaches. We need students to recognize that, while most may accept that there are sustainability issues as apparent in scientific evidence, there may be difference in how problems should be tackled. There is a belief that "through a process of rational discussion and argumentation in which everyone involved conforms to certain procedural requirements, the best arguments will gain, if not the approval, at least the acceptance of most participants" (Ojala 2013, p. 167). However, Ojala contends that emotions are inherent in and influence this deliberative process and should not be dismissed as irrational forces. Emotions and values are linked and are the forces that drive individuals to engage in the deliberative process. Emotions open up values and thus are not to be sidelined in ESD thinking. Further,

> Just because a pupil reacts emotionally in a spontaneous way, for instance, when deliberating about sustainability issues, does not mean that these emotions are non-intentional in a deeper sense. They can be based on well deliberated values, either at a personal or a cultural level, which have been stored in memory and now are being re-actualized in a specific situation, values that can be important to articulate and to discuss in the educational setting (p. 168).

The rational and the emotional need equal weight in the educational process Pigozzi (2010) notes that these foci on open endedness, deliberation, emotion, etc have implications for how ESD is positioned within schooling and with the broader education of the young person. Kanaujia, one of the authors of this chapter, was involved in designing a species focused conservation education programme for the Gangetic Dolphin along river Ganges in India. It was reckoned at the inception that the emotional bond with the dolphin had to be brought for the education programme to engage students in the topic.

The students learnt about the river ecosystem, the dolphin's morphology and physiology, its habits and very much about its threats and its importance for the health of the river itself. One of the learning materials had stories from the folklore about the Gangetic dolphin. Importantly, it was stories about the dolphin that served as the bridge to reach the students.

Often there is a discrepancy in schools between the aspirations of changing towards problem solving values laden action oriented focus and practices of knowledge/awareness acquisition (Stevenson 2007). The complex interaction of non-formal and informal communities, on the one hand, with formal schooling, on the other is a crucial component in shaping the informants' being in the world. As Payne (2005) suggests, a broader focus than the contrived school-based 'intervention' is helpful if we want to understand the effects of ESD.

> ….what the role of the formal school should be in the learning process since this is dependent on and interrelated to what proceeds in informal and non-formal settings, which can vary extensively between individuals and contexts. Should school, for instance, be the place to introduce and discuss a core of values of intergenerational and global responsibility, and offer the students contrasting perspectives on this core of values and on a range of solutions and action strategies? Should school open up for other sustainability learning forums for

> children and young people, as a complement to formal education, in order to stimulate action-experience in self chosen belongingness? (Almers 2013, p. 125)

Change requires commitment by society and by government institutions but the establishment is slow to change as innovation is contested by the powers of inertia and inherent conservatism. Change in the focus of education is a political process. Frequently, institutions and governments only change when change is essential for their credibility within their population. Education is no different; it is a conservative institution and change seldom comes from within education alone. This is not to say that educational institutions cannot or will not change but rather it requires a mood for reform within the population behind deliberations. There is something of a 'chicken and egg' dilemma which is broken only by political action. Today, however, the neoliberal political agenda in many countries across the world has encouraged schooling to pull back from complexity and return to the basics and traditional disciplinary study which explicit values questions of what society *should do* are largely ignored being seen as too controversial or too risky for society. There is a tendency to implement first order changes in policy statements, programmes and the like without adopting second order changes which impact on the fundamental goals and roles of educational institutions. Moving towards the latter is the challenge for ESD in South Asia. Moreover the challenges of sustainability and educating for sustainability will likely differ across cultural/geographical contexts between and within countries and take on different pathways and identities. The sort of education and the approaches used that are most effective in an affluent urban district are unlikely to be the same as those most useful in an impoverished rural area. It has to be accepted that there are dangers in a piecemeal approach that may result. At the heart of ESD, however, needs to be the sort of world we seek to pass to future generations. The challenge is to create an education system with the capacity and confidence to implement change and the evidence base beyond the anecdotal of how education, formally and informally, can act as a lever of change. One can but agree with Sarabhai (2015) that, "Only then will education be taken seriously as a part of the strategy for achieving (sustainability) goals" (p. 123).

Acknowledgments The authors wish to acknowledge the assistance of Dr. Philip Stimpson in the preparation of this chapter.

References

Almers, E. (2013). Pathways to action competence for sustainability—Six themes. *The Journal of Environmental Education, 44*(2), 116–127.

Delors, J. (1996). *Learning: The treasure within. Report to UNESCO of the international commission on education for the twenty-first century.* http://unesdoc.unesco.org/images/0010/001095/109590eo.pdf. Accessed 29 Aug 2015.

Dickson, J. L., Crain, R., Yalowitz, S., & Cherry, T. M. (2013). How framing climate change influences citizen scientists' intentions to do something about it. *Journal of Environmental Education, 44*(3), 145–158.

Economic Commission for Europe. (2011). *Learning for the future: Competences in education for sustainable development. Report of the committee on environmental policy* (United Nations Economic Commission for Europe Steering Committee on Education for *challenges*). Berlin: Springer.

Gadotti, M. (2010). Reorienting education practices towards sustainability. *Journal of Education for Sustainable Development, 4*(2), 203–211.

Gandhi, I. (1972). *The unfinished revolution, speech of Mrs. Indira Gandhi duriing Stockholm conference, Sweden in 1972*. Bulletin of the Atomic Scientists.

Goralnik, L., & Nelson, M. P. (2011). Framing a philosophy of environmental action: Aldo Leopold, John Muir, and the importance of community. *The Journal of Environmental Education, 42*(3), 181–192.

Hogan, D., & Tormey, R. (2008). A perspective on the relationship between development education and education for sustainable development. In *Policy and practice: A development education review*. http://www.developmenteducationreview.com. Accessed 20 Aug 2015.

Kopnina, H. (2012). Education for sustainable development (ESD): The turn away from 'environment' in environmental education? *Environmental Education Research, 18*(5), 699–717.

Lawale, S., & Bory-adams, A. (2010). The decade of education for sustainable development: Towards four pillars of learning. *Development, 53*(4), 547–550.

Lee, J. C. K., & Efird, R. (2014). *Introduction in schooling for sustainable development across the Pacific*. Dordrecht: Springer.

Mahruf, M., & Shohel, C. (2011). Models of education for sustainable development and nonformal primary education in Bangladesh. *Journal of Education for Sustainable Development, 5*(1), 129–139.

McKensie, M., Bieler, M., & McNeil, R. (2015). Education policy mobility: Reimagining sustainability in neoliberal times. *Environmental Education Research, 21*(3), 319–337.

Mogensen, F., & Schnack, K. (2010). The action competence approach and the 'new' discourses of education for sustainable development, competence and quality criteria. *Environmental Education Research, 16*(1), 59–74.

Murraya, P., Goodhewb, J., & Murraya, S. (2014). The heart of ESD: Personally engaging learners with sustainability. *Environmental Education Research, 20*(5), 718–734.

Nambiar, P. (2014). Developing learning opportunities for sustainability. *Journal of Education for Sustainable Development, 8*, 87–88.

Nambiar, P., & Sarabhai, K. V. (2015). Challenges that lie ahead for ESD. *Journal of Education for Sustainable Development, 9*(1), 1–3.

Noguchi, F., Guevara, J. R., & Yorozu, R. (2015). *Communities in action lifelong learning for sustainable development united nations educational, scientific and cultural organization*. Hamburg: UNESCO.

Ojala, M. (2013). Emotional Awareness: *On the importance of including emotional aspects in education for sustainable development (ESD). Education for Sustainable Development, 7*, 167–182.

Pandya, M., Gorana, N. R. (2011). *Paryavaran Mitra, teacher's handbook*. Ahmadabad: CEE.

Pavlova, M. (2013). Teaching and learning for sustainable development: ESD research in technology education. *International Journal of Technology and Design Education, 23*(3), 733–748.

Payne, P. G. (2005). Lifeworld and textualism: Reassembling the researcher/ed and "others.". *Environmental Education Research, 11*(4).

Pigozzi, M. J. (2010). Implementing the UN decade of education for sustainable development (DESD): Achievements, open questions and strategies for the way forward. *International Review of Education, 56*(2–3), 255–269.

Sarabhai, K. V. (2005). *Education for sustainable future*. http://www.ceeindia.org/esf/esf.asp Accessed 15 Dec 2015.

Sarabhai, K. (2010). ESD in a developing nation. *Tomorrow Today*. Paris: UNESCO (Published by Tudor Rose on behalf of UNESCO).

Sarabhai, K. V. (2015). ESD for sustainable development goals (SDGs). *Journal of Education for Sustainable Development, 9*(2), 121–123.

Sarabhai, V. K., Pandya, M., Namagiri, R. (2007). *Tbilisi to Ahmedabad. The journey of environmental education. A source book.* Ahmadabad: CEE.

Schnack, K. (1994). Some further comments on the action competence debate. In B. B. Jensen & K. Schnack (Eds.), *Action competence as key concepts in critical pedagogy.* Copenhagen: Royal Danish School of Educational Studies.

Schnack, K. (2000). Action competence as a curriculum perspective. In B. B. Jensen, K. Schnack, & V. Simovska (Eds.), *Critical environmental and health education: Research issues and challenges* (pp. 107–126). Copenhagen: The Research Center for Environmental and Health Education, The Danish University of Education.

Schusler, T. M., Krasny, M. E., Peters, S. J., & Decker, D. J. (2009). Developing citizens and communities through youth environmental action. *Environmental Education Research, 15*(1), 111–127.

Stevenson, R. B. (2007). Schooling and environmental education: Contradictions in purpose and practice. *Environmental Education Research, 13*(2), 139–153.

Sund, L., & Ohman, J. (2014). On the need to repoliticise environmental and sustainability education: Rethinking the postpolitical consensus. *Environmental Education Research, 20*(5), 639–659.

Sustainable Developmengt Goals (SDGs). http://www.un.org/ga/search/view_doc.asp?symbol=A/RES/70/1&Lang=E. Accessed 10 Dec 2015.

Taylor, P., Littledyke, M., Eames, C., & Coll, K. R. (2009). *Environmental education in context, an international perspective on the development of environmental education.* Rotterdam: Sense Publishers.

United Nations Conference on the Human Environment. (1972). *Rio+20 UNCSD.* http://www.uncsd2012.org/history.html#sthash.1TQMWb.dpuf. Accessed 10 Dec 2015.

United Nations Environment Programme UNEP. (2001). Annual report.

United Nations Organization for Education, Science and Culture UNESCO. (2004). *Contributing to a more sustainable future: Quality education, life skills and education for sustainable development.* Retrieved August 20, 2015, from http://unesdoc.unesco.org/images/0014/001410/141019e.pdf

United Nations Organization for Education, Science and Culture UNESCO. (2009). *Bonn declaration on education for sustainable development.* Retrieved August 20, 2015, from http://unesdoc.unesco.org/images/0018/001887/188799e.pdf

United Nations Organization for Education, Science and Culture UNESCO. (2010). *Tomorrow Today*, Published by Tudor Rose on behalf of UNESCO.

Willow, A. J. (2015). The shifting topology of environmentalism: Human-environment relationships and conceptual trends in two North American organizational histories. *Nature and Culture, 10*(2), 157–177.

Chapter 3
The Decade of Education for Sustainable Development in South Asia

Madhavi Joshi and Ramboojh Yadav

3.1 Introduction

"Everyone has a right to Education" asserts the Universal Declaration of Human Rights 1948. Why should education be a human right? What justifies such a right? "We are free only if we know so in proportion to our knowledge. There is no freedom without choice, and there is no choice without knowledge—or none that is illusory" (Cardozo 2000: 104). It is education which bestows such freedom of mind. This explains why education is deemed a fundamental right.

The Declaration from the World Conference on Education for All held in 1990 in Jomtien, Thailand, states, "Education can help ensure a safer, healthier, more prosperous and environmentally sound development, while simultaneously contributing to social, economic, and cultural progress, tolerance and international cooperation"… and is an indispensable key to (though not a sufficient condition for) personal and social improvement (http://www.unesco.org/education/wef/en-conf/Jomtien%20Declaration%20eng.shtm. Accessed on May 1, 2014).

"Literacy is a means for development, enabling people to access new opportunities and learn new patterns of behaviour consistent with sustainability. Literacy is also a fundamental means of participating in knowledge societies, where accessing and processing text-based information are vital" (Wals 2009: 11). The United Nations Literacy Decade (2003–2012) reaffirms the central role of literacy in development. "For many years, the international community has linked literacy and

M. Joshi (✉)
Programme Director, Youth Programmes, Waste Management Secretariat Coordinator, DESD India, Centre for Environment Education (CEE), Ahmedabad, India
e-mail: madhavi.joshi@ceeindia.org

R. Yadav
Section Chief & Programme Officer for Natural Sciences, UNESCO, New Delhi, India
e-mail: r.boojh@unesco.org

R.N. Gorana, P.R. Kanaujia (eds.), *Reorienting Educational Efforts for Sustainable Development*, Schooling for Sustainable Development,
DOI 10.1007/978-94-017-7622-6_3

development. Literacy rates are a component of the Human Development Index, and the distribution of illiteracy in the world corresponds closely with that of poverty. Literacy is an indicator of quality of schooling, access to information and, more generally, of the capacity of individuals and communities for autonomous learning, for the initiation and management of change and for active participation in society" (Wals 2009: 10).

"The right to literacy supports the pursuit of other human rights. In other words, literacy has the potential to enhance people's ability to act in the pursuit of freedom … and to empower them to interpret and transform their realities…Strong literacy skills are associated with a range of valuable and desirable outcomes (St. Clair 2010). The question, then, is not so much what literacy can do for people, but rather what people can do with literacy. How it is acquired and how it is used determines its value for the learner." (UNICEF 2013: 17). However, it must be remembered that education and learning are part of the iterative dynamic of social change: in order to change society, we need to change the way we learn and educate, and in order to change the way we learn and educate we need to change society (Wade and Parker 2008: 3).

The importance of education and learning was recognized in the discourses of human development, and then equally by the discourses of environment and sustainable development. The discourse of environmental education (EE) recognized the importance of education for Environment and Sustainable Development. Education was recognized as an important tool to promote the protection of environment and, later was acknowledged as a pre-requisite for development.

Pressure for more sustainable forms of development grew out of social movements concerned about damage to the bio-physical environment and the extent of world poverty, originating in the 1960s. Early tensions between these movements (development is needed to lift people out of poverty yet it damages the environment) were addressed by the World Commission on the Environment and Development in the 1980s. Subsequent Conferences on environment and development, (termed Earth Summits, in 1992 and 2002 led to related declarations and conventions, on such issues as biodiversity and climate change, together with action plans at international, national and local levels (Huckle 2012: 2). Key references during the UN Conference on Human Environment in Stockholm in 1972, to a chapter dedicated to Education and Training in Agenda 21 of the United Nations Conference on Environment and Development in Rio de' Janeiro in 1992, it became increasingly clear that sustainability cannot be achieved without giving due importance to education.

The 1972 International Conference on Human Environment, known as the Stockholm Conference, was the first global recognition that the environment was endangered and that a collective effort was required to address it. Principle 19 of the Stockholm Declaration states, "Education in environmental matters, for the younger generation, as well as adults, giving due consideration to the underprivileged, is essential in order to broaden the basis for an enlightened opinion and responsible conduct by individuals, enterprises and communities in protecting and improving the environment in its full human dimension" (http://www.un-documents.net/unchedec.htm. Accessed on May 1, 2014). The Stockholm Conference may be seen as the precursor to redefining the intent of education for environment and development.

"Education has risen higher on the international and national agendas in the last two decades since Jomtien (1990) highlighted the importance and relevance of education for development, closely followed by the Earth Summit in 1992, which highlighted the role of education in sustainable development" (Wade and Parker 2008: 3). The UN Conference on Education and Development (UNCED) held in Rio in June 1992 was a watershed conference which made it plain that we can no longer think of environment and economic and social development as isolated fields. Maurice Strong, Secretary-General to UNCED said, "At the core of this shift there is a need for fundamental change. Change to our economic life, a more careful and more caring use of the Earth's resources and greater co-operation and equity in sharing the benefits as well as the risks of our technological civilisation. Of particular importance is the need to integrate the ecological dimension into education and culture as well as in economics" (Palmer 2002: 69).

The Conference produced Agenda 21, a blueprint of a work programme aimed at the twin requirements of a high quality environment and a healthy economy for all people of the world. Conferences like these actually broadened the base of education and raised its profile as an instrument necessary for meeting the new requirement of development (Keating 1993). As a first, the Commission on Sustainable Development (CSD) was created in December 1992, to ensure effective follow-up of UNCED and to monitor and report on implementation of the agreements at international, regional, national and local levels and later on providing policy guidance to follow up on the Johannesburg Plan of Implementation (JPOI).

UNESCO (1997) prepared a discussion paper for the second in the series of intergovernmental conferences on Environment Education held in Thessaloniki, Greece, in 1997. This discussion paper, titled *Educating for a Sustainable Future: A Transdisciplinary Vision for Concerted Action*, stated:

> It is widely agreed that education is the most effective means that society possess for confronting the challenges of the future… Education, to be certain is not the whole answer to every problem. But education, in its broadest sense, must be a vital part of all efforts to imagine and create new relations among people and to foster greater respect for the needs of the environment… Education is also the means for disseminating knowledge and developing skills, for bringing about desired changes in behaviours, values and lifestyles, and for promoting public support for the continuing and fundamental changes that will be required if humanity is to alter its course, leaving the familiar path that is leading towards growing difficulties and possible catastrophe, and starting the uphill climb towards sustainability. Education, in short, is humanity's best hope and most effective means in the quest to achieve sustainable development (http://www.unesco.org/education/tlsf/mods/theme_a/popups/mod01t05s01.html. Accessed on May 1, 2014).

Education for Sustainable Development (ESD) "emerged in the 1990s, shaped to a great extent by the discourses of practitioners of environmental education and development education" (Huckle 2012: 4).

The Ahmedabad Declaration 2007: A Call to Action; Education for Life: Life through Education; 28th November 2007, from the International Conference on Environmental Education, held at the Centre for Environment Education,

Ahmedabad, India states, "Environmental Education processes support and champion Education for Sustainable Development. Such education processes must be relevant, responsive and accountable…We pledge to build partnerships and share our diverse experiences and collective knowledge to refine the vision of sustainability while continually expanding its practice" (CEE 2007) (http://www.unevoc.net/fileadmin/user_upload/docs/AhmedabadDeclaration.pdf. Accessed on May 1, 2014).

3.2 The Decade of Education for Sustainable Development

"The UN Decade of Education for Sustainable Development (DESD, 2005–2014), offers an opportunity to rethink the manner in which we approach global challenges" (Wals 2009: 3). "The creation of the DESD was seen as a means to re-emphasize these overarching goals in a context of sustainable development by emphasizing the role of education and learning. Resolution 57/254 on the United Nations Decade of Education for Sustainable Development (2005–2014) was adopted by the United Nations General Assembly in December 2002, shortly after the World Summit on Sustainable Development (Rio plus 10) which was held in Johannesburg in August/September of the same year. This resolution requested UNESCO, as the designated lead agency for overseeing the Decade, to develop an International Implementation Scheme (IIS) which would also clarify the relationship of the DESD with existing UN-supported educational processes (e.g., the Dakar Framework for Action and the United Nations Literacy Decade).

The IIS was prepared through consultations with other UN agencies, international organizations, governments, NGOs, universities, individuals and the private sector, building upon more than 2000 contributions gathered through the recommendations of national, regional and global workshops/conferences and through the UNESCO-DESD website (Wals 2009: 8).

The final version of the IIS was presented and approved by the United Nations General Assembly in September 2005. The basic vision of the Decade was of a world in which everyone had the opportunity to benefit from education and learns the values, behaviours and lifestyles required for a sustainable future and for positive societal transformation. DESD sought to promote the meaningful development and implementation of ESD on all geographical scales (locally, nationally, regionally and internationally) with the involvement of a wide range of stakeholders (Wals 2009: 8). The Decade as envisaged would further strengthen the implementation of the Millennium Development Goals, the Education for All Goals and the UN Decade of Literacy by countries.

The International Implementation Scheme was developed by UNESCO in consultation with UNEP; intergovernmental organizations; UNESCO Member States involving: the broader UN community; non-governmental organizations; universities and partner networks; and reviews by leading academics and experts in the field providing a framework for all partners to contribute to the Decade.

Education's role and agenda for empowerment and in particular action-orientation (to education) was a key outcome from the 'Education for a Sustainable Future' Conference held in India by the Centre for Environment Education (CEE) in January 2005, before the UN DESD was formally launched. Key inputs from the over 900 participants, most of them from civil society organizations from 60 countries, informed the reflections about the Decade's road ahead. The Ahmedabad Declaration, made at this conference, reiterated that sustainable development is the empowerment of people and that the key to such empowerment is action-oriented education recognizing that we are all learners as well as teachers. It stated that ESD can add "substance and vigour" to DESD (Sarabhai et al. 2007: 109).

ESD therefore, recognizes the need for a paradigm shift, not just in development but in a way in which education and learning takes place. Wals in the DESD mid-decade review titled *Review of Context and Structures for Education for Sustainable Development* notes that there is a greater consensus over the following key principles covering the scope, purpose and practice of ESD:

- "A transformative and reflective process that seeks to integrate values and perceptions of sustainability into not only education systems but one's everyday personal and professional life;
- A means of empowering people with new knowledge and skills to help resolve common issues that challenge global society's collective life now and in the future;
- A holistic approach to achieve economic and social justice and respect for all life;
- A means to improve the quality of basic education, to reorient existing educational programmes and to raise awareness" (Wals 2009: 26).

> ESD is far more than teaching knowledge and principles related to sustainability. ESD, in its broadest sense, is education for social transformation with the goal of creating more sustainable societies. ESD touches every aspect of education including planning, policy development, programme implementation, finance, curricula, teaching, learning, assessment, administration. ESD aims to provide coherent interaction between education, public awareness, and training with a view to creating a more sustainable future (UNESCO 2012: 33)

"The Bonn Declaration gave the world an action plan for ESD and outlined steps for implementing the remainder of the Decade. The knowledge, technology and skills already exist to turn around unsustainable development models, the declaration said. It is now imperative to act to bring about long-term change" (Nolan 2012: 11). The year 2009 marked the mid-way point of the Decade with the organizing of the 'UNESCO World Conference on Education for Sustainable Development – Moving into the Second Half of the Decade' in Bonn, Germany. The Bonn Conference, attended by about 50 education ministers and vice-ministers, took stock of the progress made so far and recommended aligning current strategies with emerging concerns and challenges. The report from the Bonn Conference further states that:

> There is increased recognition that this challenge (for sustainable development) cannot be solved only through technological advances, legislative measures and new policy frameworks…while such responses are necessary, they will need to be accompanied by changes in mindsets, values and lifestyles, as well as a strengthening of people's capacities to bring about change. This recognition explains the key role many governments, NGOs, UN agencies and indeed, companies are allocating to learning and capacity-building as they search for solutions to interrelated sustainability challenges such as climate change, disaster risk management, biodiversity loss, sustainable production and consumption (Wals 2012: 9).

3.3 Strategies in the Region

"Although both the challenge of sustainable development and the call for ESD is worldwide, there is a general understanding that the local realities and manifestations of 'unsustainability' are often quite different and deeply rooted in local histories and political and cultural traditions. Therefore, regional strategies for the development and implementation of ESD have been developed…" for each region (Wals and Kieft 2010: 7). At the beginning of 2005, UNESCO Asia and the Pacific Regional Bureau commissioned a Situational Analysis of DESD in the region with financial support from the Japanese Funds-in-Trust. The Situational Analysis carried out in all the sub-regions of Asia and the Pacific, including South Asia, provided a snapshot of ESD initiatives and highlights various opportunities for ESD in the region. This exercise guided the drafting, through a consultative process, of the Asia Pacific Regional Strategy for DESD. This strategy was launched in Nagoya, Japan, in 2005. The discussion and analysis during the formulation of the strategy clearly pointed towards the importance of highlighting the unique and diverse situations in the region. These would at the same time challenge and be an opportunity for developing contextual responses. It recognized the importance of tradition and culture, and the strong role of youth, rural and indigenous communities in ESD.

The UNESCO Asia and the Pacific Regional Bureau conducted sub-regional coordination and capacity building workshops and consultations in 2008–2009 involving the UNESCO National Commissions, and partners. Key findings from these consultations were presented at the 2009 mid-decade UNESCO Conference in Bonn. The findings point to the need to integrate ESD as a part of the National Sustainable Development strategies, budgetary allocation, inter-sector synergy and coordination, and support for capacity building. The consultations suggested that:

> ESD in the Asia-Pacific region is now at a point where countries are beginning to engage in clear strategies for ESD – moving from ESD in theory to practice – by identifying clear thematic national SD priorities, linking priorities to existing aims and objectives for education and learning in current policies, building inter-ministerial support, discussion financing and engaging with the right people at the national level. These recent developments are steering towards a goal to incorporate or link ESD to national development plans, national sustainable development strategies and/or poverty reduction strategies to increase the importance of the role of ESD in addressing the growing list of recent sustainable development challenges in the Asia-Pacific region (Wals and Kieft 2010: 19).

3.4 Environment and Development in the South Asia Context

South Asia is the poorest region on the earth after Sub-Saharan Africa and has the lowest GDP per capita.

> Poverty, population pressures, natural resource degradation and depletion remain a challenge despite achievements in areas of literacy, child immunization and population control in countries like Sri Lanka and the Maldives. Rapid economic and social development over the last 15 years has been remarkable, but enormous challenges remain for improving the quality of life as more than 40 percent of the population continues to subsist on less than one US dollar a day (Joshi and Jain 2005: 55).

Regional and national assessments prioritize the need to eliminate poverty and create human security, manage population growth, conserve natural resources and create a secure economic base. Food and income security are also key priorities. The high population growth rate of 1.8 % puts pressure on land, water and other natural resources. Joshi and Jain in the Sub-regional overview of South Asia reported that:

> Natural resource depletion and ecological degradation pose serious long-term sustainability concerns. Significant natural resource concerns include water quality degradation and local and regional water scarcity; dwindling forests, coastal wetlands, freshwater bodies and fisheries; soil degradation due to nutrient depletion and salinization; and poorly managed water resources. Many of these issues are particularly important for the rural poor and disadvantaged groups. The causes of resource depletion and environmental degradation are many – increasing urban lifestyles; overpopulation; inadequate legal and policy frameworks for protection; subsidies and policy distortions; and limited data on resources and environmental quality (2005: 55).

The sub-region is increasingly vulnerable to the impacts of natural disasters and climate change. Also, of critical importance are the socio-political realities of the countries where conflicts and fragile political systems in most countries pose challenges to a sustainable future.

3.5 Decade of Education for Sustainable Development in South Asia: Key Actions

The launch of the Decade in 2005 had a gradual resonance in the countries in South Asia. The IUCN office in Pakistan organized an ESD workshop and translated the UN DESD International Implementation Scheme (IIS) into Urdu. The Indian National Commission for Cooperation with UNESCO identified the Centre for Environment Education as the Nodal Implementing Agency for DESD in India. Sri Lanka and Maldives integrated ESD into their thematic focus on Peace and Climate Change respectively.

The implementation and operationalization of ESD in the South Asian countries have depended on the regional, national and local differences which have influenced the meaning and context of ESD as well. However, the interconnections between the economic, environmental, social and cultural dimensions of development are the common thread that runs through all education and learning for sustainable development. Traditionally environmental protection and conservation has been a part of life of the people living in this part of the world. In that context, the definition of 'environment' has been fairly broad, closer to how 'sustainability' is understood today. ESD has helped provide the focus on a paradigm shift to 'action-orientedness' and 'change' to be brought about by education.

3.6 ESD and Policy: Institutional Mechanisms

Almost all the countries of the sub-region have implemented ESD in some form and have set up a national ESD coordination body or a National Focal Point in the Ministries of Education. India has also designated a nodal institution, CEE, as the National implementing agency for DESD. In Pakistan, UNESCO and IUCN together with the Ministries of Environment and Education formed an informal core group to implement and promote ESD activities in the country. Bhutan's Gross National Happiness (GNH) Index largely informs an understanding of ESD in that country which has set up an independent GNH Commission in the Ministry of Planning to inform all polices and plan of the country. The UNDP–Bhutan Development Report (UNDP 2000) "…highlights and clarifies how Bhutan is choosing its own path in order to avoid mistakes that have characterized the development of other regions of the world. GNH calls for a balanced, multi-dimensional approach to development, harmonizing environmental, cultural and spiritual values, good governance, and economic growth. It is important to notice that the concept of GNH has evolved as a reaction to the experiences of other developing countries" (Ezechieli 2003: 23). As the Report states, "Bhutan's late start in development has had one major advantage: It allowed the country to learn from the experience of others. The pursuit of growth in GNH rather than in Gross National Product (GNP) reflects Bhutan's anxiety to avoid some of the more glaring failures of the blind pursuit of economic growth" (UNDP 2000: 17).

However overall, there appears to be a lack of a structure that facilitates interdepartmental governmental cooperation in ESD. This is possibly due to insufficient experience in cross-boundary thinking among policy-makers. Nonetheless, many countries are showing increasing efforts to establish ESD linkages between ministries and other stakeholders.

There is a notable presence of ESD in a wide range of national policy documents such as Five Year Plans, the National Conservation Strategy and increasing understanding for connecting national sustainability goals to ESD. In most cases ESD is integrated into national educational policies and curricula at the primary and secondary levels of education, building on work being done in EE. ESD also appears in national environmental and sustainable development policies. For example, vulnerability to climate change and the need for disaster risk reduction and peace have influenced ESD policy development in the countries of this region. The development strategy of the Republic of Maldives, its priorities, goals and path to development has been set out in the 7th National Development Plan (NDP) (Planning Commission 2007) (http://www.planning.gov.mv/en/images/stories/ndp/seventh_ndp.pdf. Accessed May 1, 2014). This Plan presents an analysis of and context within which issues exist in the Maldives and suggests policies, strategies and interventions to overcome the issues. The Safe Island strategy of the government has identified three main areas of cooperation, namely, social and economic equity; environmental management and disaster risk reduction; and governance. The government is also promoting programmes such as environment and risk management through participatory and community-based approaches that reinforce environmental protection in the areas of erosion, waste management and energy, with a focus on vulnerable groups and youth.

The presence of specific ESD policies is not common in most countries. To this end, an international and regional impetus is needed to encourage countries to develop ESD-specific policies and strategies and to ensure their implementation.

3.7 ESD in Formal Education

The UN General Assembly resolution (59/237) for creating the DESD encouraged Governments to consider the inclusion of measures to implement the Decade in their respective education systems and national development plans. The inclusion of ESD in formal education, especially at the primary and secondary levels, indicates progress in most countries. All countries in the region had initiated or had in place EE policies and initiatives to integrate EE in formal education when DESD was announced. This provided a foundation to broaden the context and integrate ESD in formal education systems. India's National Curriculum Framework (NCF), 2005, embodies the integration and values-based orientation for ESD. The National Conservation Strategy of Pakistan formed the basis for ESD in the country. Sri Lanka included peace as a key theme to focus on ESD in the country. The Education Development Centre in the Maldives has developed the National Curriculum Framework for the country. Being vulnerable to climate change, Maldives has a specific focus on disaster risk reduction.

ESD was aligned with the efforts being made towards achieving targets for the Millennium Development Goals (MDGs), United Nations Literacy Decade (UNLD) and the Education for All (EFA). The EFA goals call for the provision of quality education for all stages of the life cycle, from early childhood (Goal 1), to primary schools age (Goal 2) to youth and adult skills and education (Goals 3 and 4) both in formal and non-formal settings. Goals 5 and 6 ensure gender and social equality in terms of access to quality education. Hence the EFA goals call for lifelong education programmes with a focus on quantity ("all"), equality and quality. Between 2000 and 2009 the number of children enrolled in pre-primary education showed a significant increase in South Asia. Bangladesh, Bhutan, India, Maldives, Nepal,

Pakistan and Sri Lanka have policies, guidelines and frameworks in place for Early Childhood Care Education (ECCE). Improving the net enrolment ratio in primary education has been on track especially in India, Maldives and Sri Lanka. The gross enrolment ratio in secondary education needs much improvement overall except in Maldives and Sri Lanka. Gender disparity and fewer enrolments in rural areas compared to urban areas is still a concern for the countries in this region (Jo-Kim 2012). While the access to education goal is set to be achieved in most countries in South Asia, the quality of education has to be the focus in the years ahead.

Inclusion of sustainability issues in existing curricula and redesigning curricula are two of the ways in which countries have integrated ESD in formal education settings. The more recent initiative by CEE in India is the *Paryavaran Mitra* (Friends of Environment) programme involving 150,000 schools in sustainability action. Methodologies that inculcate critical thinking are evolving and need to be developed further. "The Bhutanese schools also focus on mindfulness and care/compassion for the learners. So there is an attempt to address all abilities within an inclusive school environment (GMES, Bhutan)" (Nolan 2012: 27). The development and inclusion of SD or ESD in curricula seems to be causing a co-evolution of pedagogy. ESD's presence "goes hand-in-hand with a rethinking of the kind of learning necessary to address sustainability issues. ESD is thus becoming a catalyst of educational reform" (Wals 2012: 32). Despite these examples, "Currently, the general picture of formal education is that it does not yet point to a major reform of the educational systems to better incorporate sustainability and contribute to sustainable development but rather minor adjustments to the existing system" (Wals and Kieft 2010: 8).

Much needs to be done to integrate ESD in higher education institutions. While EE has been integrated to a great extent in the school system, it is still quite peripheral in higher education institutions. For example, in India, EE had not received adequate attention in higher education until the Supreme Court Directive in 2001. The Supreme Court directed the University Grants Commission (UGC) to introduce a basic course on the environment at every level in higher education.

3.8 Capacity Building Teachers and Educators

ESD requires reorienting teaching and learning methodologies and resources. Efforts have been made to integrate ESD into existing teacher training programmes in the different countries. In Bhutan, workshops have been held for capacity building teachers for GNH and the development of an action plan and indicators to support implementation of ESD, by the Ministry of Education and UNESCO. The school ESD plan for Bhutan includes individual health, social and cultural well-being and environment as key areas to support capacity building and ESD action in schools. *Maldives Vision 2020* document highlights enhancement of training opportunities for local teachers as one of the major strategies to improve access as well as quality of education. (http://planipolis.iiep.unesco.org/upload/Maldives/

Maldives_NPA%20EFA.pdf. Accessed on May 1, 2014). UNESCO New Delhi is working with the Maldives Government to implement an ESD programme focusing on climate change education, with a focus on teacher education. Maldives is endeavouring to build capacity of local teachers to address the challenge it faces in the availability of trained teachers from the country itself. In Sri Lanka, UNESCO is implementing island-wide teacher training in ESD basically focusing on education for peace and sustainable development. CEE in India has developed teacher training programmes and resource material to enable teachers to engage with sustainability teaching and learning. These have included approaches and methodologies that foster critical and out-of-the-box thinking and understanding complex connections and interrelatedness of various issues. Teacher training remains a challenging task for ESD today and in future in spite of the recognition ~~to~~ of the need to build capacity. It is important for the relevant agencies to look at ways of making available adequate resources, both financial and mentoring, for this to happen. All the countries have also been laying a great emphasis on locale-specificity and the need to use local languages in educational resource materials. In Sri Lanka, in recent years Tamil and Sinhala have been offered in schools particularly for achieving better social cohesion and peace and these interventions are clearly included and reflected in the National Action Plan on Education for Peace and Sustainable Development (EPSD) launched in January 2012. Special teacher training programmes for those who were teaching Tamil or Sinhala as a second language have been organized.

3.9 ESD in Technical and Vocational Education

"Technical and Vocational Education and Training (TVET), or education for the world of work, provides learning and life-skills programmes for young people and adults. TVET is essential for the expansion of skills and development of competencies necessary in rapidly changing labour markets" (Nolan 2012: 46). This kind of education is seen as one of the solutions to poverty reduction and a support to socio-economic development. Besides technical knowledge and the capacity for productive team work, people now have a preparation that goes beyond basic literacy and numeracy to include both vocational and social skills, together with values that help build harmonious societies. Increasingly, this approach to education for the world of work is considered a lifelong learning programme that takes place not only in schools but also in the workplace.

"Seen through the lens of TVET, ESD is a requisite to ensure sustainable livelihoods and occupations. Integrating ESD into TVET is essential to develop knowledge and skills that support economic development and also enable people to improve the quality of their daily lives" (Nolan 2012: 46). South Asia, having the highest number of young people in the world, has to increasingly enhance technical and vocational education opportunities while developing formal school and higher education systems. All the countries already have in place major programmes focusing on technical and vocational education and integrating ESD is recognized as being a strengthening factor. For example, the Bhutanese delegation to the South Asia ESD Capacity Building Programme, organised by the UNESCO Asia and the Pacific Bureau, in 2008, felt that already rich experience in EE, vocational orientation and values education would provide a starting point for initiating and strengthening ESD in the country.

Greater emphasis has been put on TVET in national education policies due to expanding economic development in the region as well as changes in labour markets. Several countries have accepted TVET as a means to equip students with the skills and knowledge they need to meet the demands of a rapidly changing economy.

> In many countries enrolment in TVET is limited to a small group. Women in particular often lack opportunities to enter TVET. Although data on female enrolment in TVET are often limited, available data from the region show that in most countries girls are under-represented and encounter barriers to accessing technical and vocational training. The disadvantages are striking when set against the global average female participation rate in TVET for 2009 of 45 per cent. The exclusion of girls from TVET hampers their chances of finding gainful employment and equipping themselves with the necessary life skills that will give them financial independence and equal opportunities for professional development (UNESCO and UNICEF 2012: 20).

3.10 ESD in Non-formal Education and Informal Learning

> Perhaps most ESD activity around the world is generated not by formal government organizations but rather by NGOs who sometimes work with formal education systems but more often work in non-formal and informal learning settings. Informal and non-formal education tends to refer to the collective learning that takes place outside of formal education systems in everyday life for instance in the context of families, work places, clubs, web-based communities, etc (Wals 2009: 54).

There are several examples of environmental education, health education and other sectors engaging formal education systems through a non-formal mode such as the eco-club. These have been used for promoting ESD values by NGOs, UN agencies and governments in these countries. Such initiatives provide important understandings in terms of different methodologies for hands-on, action-oriented learning for the participants in this process.

The CEE, for example, has involved college youth through eco-clubs in whole-institution sustainability action and, through the South Asia Youth Environment Network (SAYEN), supported by UNEP, engages youth in sustainability policy, awareness and education. UNESCO Bangladesh instituted a media award for young journalists, and supports a community radio initiative. Nepal, too, has a strong eco-club movement across the country. Corporate groups have begun engaging with sustainability education. Two such examples include the Go Eco-beyond programme supported by MAS Holdings in Sri Lanka and the Paryavaran Mitra programme supported by ArcelorMittal in India.

> Many countries developed policies and guidelines on Community Learning Centres, citing such a mechanism for providing literacy and non-formal education in their EFA national plan of action as well as other development plans and strategies (UNESCO and UNICEF 2012: 32).

The region has several initiatives that engage community-based organizations, help capacity build facilitators and educators and document good practice. Organisations such as the Asia Pacific Centre for Cultural Understanding (ACCU), Japan and ESD-J support programmes that help bring community-based learning and insight into mainstream dialogue on ESD.

3.11 Biosphere Reserves: ESD Learning Sites

Among the UNESCO instruments to promote ESD, the World Network of Biosphere Reserves plays an important role in providing demonstration and learning sites for ecological and cultural diversity and ecosystem services for human well-being. These are internationally recognized sites for environmental, economic and social (including cultural and spiritual) sustainability as well as for advancing understanding of interactions between people and nature. Biosphere Reserves are being promoted as learning laboratories and platforms for policy professionals, research and scientific communities, management practitioners and local communities to work together to translate global principles of sustainable development into locally relevant praxis. The theme workshop on 'Biosphere Reserves and World Heritage Sites: Learning Laboratories for Sustainable Development' organized as part of the 4th International Conference on Environmental Education in 2007, discussed in detail the value of these sites as areas and spaces for ESD. These include a focus on facilitating learning processes, capacity building of biosphere reserve managers, building case studies of effective collaborative management and exploring the potential utilization of information technology to manage biosphere reserves.

Non-formal education has also provided a great amount of flexibility and innovation in awareness and educational methods and materials. Understanding from the various adjectival educational experiences, such as health, peace, biodiversity, climate change, and gender, give a rich context for educators and facilitators of ESD. The broad range of methods and media includes traditional forms and also considerable information and communication technology (ICT). As Wals (2012: 26) commented, "Spaces are opening up around the world for more process-oriented, transformative approaches to ESD that require higher levels of participation and self-determination. Young people using ICT tools to make their voices heard are creating much of the momentum. The growth of social media and open source internet-based platforms is also facilitating access to education".

3.12 New Partnerships and Networking in Support of ESD

> In 2005 in Nagoya, Japan, at a conference celebrating the Asia-Pacific launch of the United Nations Decade of Education for Sustainable Development (DESD), the United Nations University (UNU) announced the formation of seven RCEs (Regional Centres of Expertise) on ESD involving collaboration and cooperation between Higher Education Institutions and other formal and non-formal educational providers. These linkages between informal, non-formal and formal education are a key element in ESD (Wals and Kieft 2010: 35).

India had ten of these RCEs out of which seven are facilitated by the CEE. The RCEs help to "…establish local/regional knowledge bases which are both culturally sensitive and relevant to their localities and in a relatively short period can establish vibrant and effective communities of practice. The pedagogic emphasis is on knowledge creation through active, contextually grounded, learning…" (Wals and Kieft 2010: 35).

In addition, the Decade of Education for Sustainable Development fostered global partnerships that explored the content and methodology of ESD. The ESD ExpertNet, one such example, is a trans-national network of experts and leaders from state and non-state institutions that is developing strategies to help enhance the implementation of Education for Sustainable Development (ESD) in the participating countries. Currently, experts from four countries are members of the network: Germany, India, Mexico and South Africa. ESD ExpertNet was earlier managed by Deutsche Gesellschaft Für Internationale Zusammenarbeit (GIZ) with core funding from the Federal Ministry for Economic Cooperation and Development (BMZ). It is now being anchored by Engagement Global (EG). As a think tank, the ESD Expert Network jointly develops and realizes innovative concepts and strategies to strengthen individual competencies and institutional capacities to implement ESD in the participating countries. The Network concentrates on three emerging economies and Germany as an industrialized country.

Since its launch in 2010, ExpertNet has been intensively working on an international core curriculum for the training of teachers and teacher trainers (ToT). The curriculum brings together content and strategies that enhance a systemic understanding of environment and development (social and economic) and offers an experience-based development of competences. The exchange of perspectives in the trans-national network allowed considering divergent needs and scoping for action in the four countries.

Based on an international core curriculum, the member institutions developed and piloted country specific versions of the ToT. These experiences have currently been evaluated in four national ESD training of multipliers handbooks. The multi-stakeholder composition of the country groups deliberately includes government institutions as well as implementing educational institutions and organizations; this facilitates the recognition and integration of developed training approaches into national education systems.

The network has also developed an international ESD leadership training programme for young professionals. This is a 1 year training cycle combining residential training phases in Germany with a 2-month internship at one of the ExpertNet partners and the planning and implementation of an innovation project at the home organization. Participants learn together to investigate and apply ESD strategies. They also develop soft skills enabling them to realize their innovation projects and to professionalize ESD provision in their home organizations. To date thirty-nine young leaders from India, Mexico, South Africa and Germany have successfully completed this intercultural training programme.

3.13 Supporting Urban Sustainability: Collaborative Learning and Ecosystem Services and Poverty Alleviation Approaches

Partnerships evolved around methodologies for engaging multi-disciplinary and multi-organisational collaborative learning. The Centre for Environment Education (CEE), the International Centre for Local Environmental Initiatives (ICLEI) and the Swedish International Centre for Education for Sustainable Development (SWEDESD) have collaborated to initiate a programme which brings together a multi-stakeholder team in cities to initiate a joint dialogue or inquiry on local ecosystem services of concern. This group explores ideas for collaborative management and restoration of ecosystem services for supporting livelihoods. The multi-stakeholder team of 3–4 persons may contain representatives from the poor, community organizations, NGOs, academic institutions and local government. The inquiry process outcomes are locale-specific and dependent on the inquiry process itself. For example, the process could result in a new policy instrument or institutional arrangement for managing a wetland, or a proposal for 'green infrastructure', such as root zone treatment, or small community actions such as those related to urban agriculture and composting. This Supporting Urban Sustainability (SUS) programme has been initiated in India, Bangladesh, Bhutan, South Africa, Tanzania, Vietnam, and Sweden.

3.14 Research and Documentation of Learnings

The United Nations Literacy Decade (UNLD), Education for All (EFA) and Millenium Development Goals (MDGs) required governments to monitor and report progress on achieving them. In the case of DESD, national mechanisms for monitoring were not set up. As the Decade did not set quantifiable or uniform goals, it was left to the individual countries themselves to develop their action plans. There was a large exercise conducted by UNESCO to gather data from all member countries and key respondents were identified for a mid-term and end-of-decade review of trends.

Some initiatives in the Asia-Pacific region, by the Japan Council on the UN Decade of Education for Sustainable Development (ESD-J) and Asia-Pacific Cultural Centre for UNESCO (ACCU) in particular have included work in South Asia as well. ESD-J conducted a 3-year documentation of best practice in ESD focusing on agrarian communities in rural areas. India was one of the participant countries. ACCU facilitated the Holistic Ownership-based Participatory Empowering (HOPE) evaluation, a participatory research project involving local communities. Bhutan, Bangladesh and Nepal were involved in this initiative. ACCU facilitated the Centres of Excellence programme where it supported programmes by the Dhaka Ahsania Mission of Bangladesh and TVE Asia Pacific of Sri Lanka. TVE Asia Pacific developed video programmes documenting ESD best practices in the Asia-Pacific region.

3.15 Key Lessons from the Decade

As the Decade ended in 2014, there were important lessons and suggestions for a way forward beyond the Decade. The following are some of the key highlights from the limited analysis for South Asia:

- Inclusion of ESD is evident in formal education, especially at primary and secondary school level, in teacher education and technical and vocational education. However there is little support for ESD in early childhood education. There is also a need for considerable progress at all levels of education to implement fully the reorientation towards ESD.
- Knowledge about ESD in informal and non-formal education still remains anecdotal, and remains an area with great learning potential for ESD practice that is beyond formal education. There is therefore a great need to create possibilities for finding ways to document, share and use this material.
- ESD research and development is a potential area especially in the context of the way higher education can contribute to ESD. Scholarships for ESD-related research, innovation and capacity building would support more academic interest in the process and learning from ESD implementation. The Mahatma Gandhi Institute of Education for Peace and Sustainable Development (MGIEP), India, a UNESCO Category I institute has been set up in New Delhi to carry out capacity building and research in education for peace and sustainable development in the Asia and the Pacific region.
- While monitoring and evaluation indicators have been developed for the Asia and the Pacific region, these are far from being implemented. Such research is required urgently to capture the process learnings from ESD implementation.
- A wide range of networks of universities, national and local governments, private sector and NGOs have been initiated, sharing their experiences and work.
- Public budgets and economic incentives for ESD are minimal or completely absent. Governments have started taking into consideration financing and budgeting for ESD across ministries by determining how ESD aligns with key

documents in different sectors to facilitate coordinated funding and inter-sectoral programming for ESD at the national level.

- A number of UN agencies have been engaged in implementing DESD in one way or another. ESD-related workshops and meetings, and numerous ESD related publications are some of the many activities and initiatives that UN agencies have undertaken in the framework of DESD implementation. However, much work remains to be done before a concerted 'delivering as one' UN response becomes a reality. In this regard, reinforcing financial and human resources could help to improve implementation of DESD.

3.16 The Ways Forward

The evolution of DESD has much to do with the concern of environmentalists and ecologists the world over for ensuring a sustainable future of the planet. The principles of sustainability have been laid down to a large extent by practicing environmental educators and scientists. The role of education has been seen as critical to bringing about change towards sustainable development. The Decade has built on the existing knowledge and expertise in education within governments, non-governmental organizations, and other expert institutions involved in environmental education, health education, peace education and many more.

ESD has become more important in view of the recent economic crisis and also ecological crises of climate change and biodiversity. The global financial and economic crisis in particular highlighted the weaknesses of prevailing economic paradigms, questioning assumptions about desirable lifestyles, development models and personal values. Across the world increasing attention is being paid to such aspects as the green economy, a desire for viable, long-term solutions and greater sustainability, especially in the perspective of other crises concerning climate change, natural disasters, food and energy. The UN Interagency Statement of 25 June 2009 on *Green Economy: a Transformation to Address Multiple Crises* states that "the shift towards a green economy requires ESD…" The economic crisis has its impact on education in terms of funds, opportunities and structures. It has also become important to think about the purpose and end objective of education in terms of what kind of citizenry education systems want to develop. There is also a need to address ESD concerns at the grassroots by making it a people's movement. ESD also needs to be translated into concrete practical actions and replicable examples.

References

Ahmedabad Declaration, A call to action. (CEE, 2007). http://www.unevoc.net/fileadmin/user_upload/docs/AhmedabadDeclaration.pdf. Accessed 1 May 2014.

Cardozo, B. (2000). *The paradoxes of legal science*. New York: Columbia University Press.

Ezechieli, E. (2003). *Beyond sustainable development: Education for gross national happiness in Bhutan*. Stanford: School of Education, Stanford University.

Huckle, J. (2012). Sustainable development. In J. Arthur & A. Peterson (Eds.), *The routledge companion to education*. Abingdon: Routledge.

Jo-Kim, G. (2012). Where do we stand on EFA and the keydevelopment gaps? *Towards EFA 2015 and beyond – Shaping a new vision of education. 9–11 May, 2012*. Thailand: UNESCO.

Joshi, M., & Jain, S. (2005). Sub regional overview. *A situational analysis of education for sustainable development in the Asia-pacific region*. Bangkok: UNESCO.

Keating, M. (1993). *The earth summit's agenda for change: A plain language version of agenda 21 and the other Rio agreements*. Geneva: The Centre for Our Common Future.

Maldives Vision. (2020). http://planipolis.iiep.unesco.org/upload/Maldives/Maldives_NPA%20 EFA.pdf. Accessed 1 May 2014.

Nolan, C. (2012). *Shaping the education of tomorrow: 2012 report on the UN decade of education for sustainable development, abridged*. Paris: UNESCO.

Palmer, J. (2002). *Environmental education in the 21st century: Theory, practice, progress and promise*. London: Routledge Falmer.

Planning Commission. (2007). National Development Plan (NDP). http://www.planning.gov.mv/en/images/stories/ndp/seventh_ndp.pdf. Accessed 1 May 2014.

Sarabhai, V. K., Pandya, M., & Namagiri, R. (2007). *Tbilisi to Ahmedabad. The journey of environmental education. A source book*. Ahmedabad: Centre for Environment Education.

St. Clair, R. (2010). *Why literacy matters: Understanding the effects of literacy education for adults*. Leicester: NIACE.

Stockholm Declaration. (1972). *Declaration of the United Nations conference on the human environment*. Stockholm 1972. http://www.un-documents.net/unchedec.htm. Accessed 1 May 2014.

UNDP. (2000). *Bhutan human development report 2000. Gross national happiness and human development: Searching for common ground*. Thimphu: United Nations Development Program Bhutan.

UNESCO. (1997). *Educating for a sustainable future: A transdisciplinary vision for concerted action*. Thessaloniki: UNESCO. http://www.unesco.org/education/tlsf/mods/theme_a/popups/mod01t05s01.html. Accessed 1 May 2014.

UNESCO. (2009). *United Nations literacy decade – International strategic framework for action*. http://unesdoc.unesco.org/images/0018/001840/184023e.pdf. Accessed 1 May 2014.

UNESCO. (2012). *Education for sustainable development – N° 4 – Sourcebook (learning & training tools)*. Paris: UNESCO.

UNESCO and UNICEF. (2012). *EFA Goal 4 – Youth and adult literacy*. Bangkok: UNESCO.

UNICEF. (2013). *Second global report on adult learning and education – Rethinking literacy*. Paris: UNESCO Institute for Lifelong Learning.

Wade, R., & Parker, J. (2008). *EFA – ESD dialogue: Educating for a sustainable world* (Education for sustainable development policy dialogue, Vol. 1). Paris: UNESCO.

Wals, A. (2009). United Nations Decade of Education for Sustainable Development (DESD, 2005–2014). *Review of context and structures for education for sustainable development. Learning for a sustainable world*. Paris: UNESCO.

Wals, A. (2012). *Shaping the education of tomorrow: 2012. Full-length report on the UN decade of education for sustainable development*. DESD monitoring and evaluation – 2012, UNESCO Education Sector. UNESCO.

Wals, A. E. J., & Kieft, G. (2010). *Education for sustainable development, research overview*, Sida Review 2010:13, Sida.

Chapter 4
Education for Sustainable Development (ESD) Initiatives and Programmes in Bangladesh

Md. Salequzzaman and Rajeswari Namagiri Gorana

4.1 Introduction

The paradigm of 'sustainable development' (SD) gained an emphatic impetus in global discussions after the 'Brundtland Commission[2] published its landmark 1987 report, *Our Common Future*. Here sustainable development is defined as development that "…meets the needs of the present generation without compromising the needs of future generations" (WCED 1987: 43). Using Brundtland's definition, Jacobs et al. (1987: 18) described the elements of sustainable development as "integration of natural resources, conservation and development; satisfaction of basic human needs; opportunities to fulfill other non-material human needs; progress towards equity and social justice; respect and support for cultural diversity; provision for social self-determination and the nurturing of self-reliance, and the maintenance of ecological integrity".

Sustainability is often used inter-changeably with sustainable development, which is a long-term dynamic process that allows individuals, organizations and societies to flourish as members of their ecological communities (Sonneborn 1998). The robustness of such development requires renewed thinking. It [sustainability] is based on the need to think long term, understand systems, recognize limits, protect nature, transform business-as-usual, practice fairness in context and fairness over time, and embrace creativity (Atkisson 1999; Heij 2001; Sutton 2001). (http://researchrepository.murdoch.edu.au/286/2/02Whole.pdf Accessed on March 29, 2012).

Md. Salequzzaman (✉)
Founding Head, Environmental Science, Khulna University, Khulna, Bangladesh
e-mail: salek_uz@yahoo.com

R.N. Gorana
Programme Coordinator, Children's Media Unit, Centre for Environment Education (CEE), Ahmedabad, India
e-mail: rajeswari.namagiri@ceeindia.org

R.N. Gorana, P.R. Kanaujia (eds.), *Reorienting Educational Efforts for Sustainable Development*, Schooling for Sustainable Development,
DOI 10.1007/978-94-017-7622-6_4

Governance, policy, technology, financial mechanisms, economic plans, rules and regulations, and education, are means for bringing about fundamental changes, but it is education most of all that can influence the other multiple means and at the same time foster such paradigmatic change. Education is the driver of change and is the driver of drivers as stated by Kartikeya Sarabhai in the international conference on Education for Sustainable Future (ESF) held in Ahmedabad in 2005 (CEE 2009: 5). The report of the UNESCO International Conference on Environment and Society; Education and Public Awareness for Sustainability held in Thessaloniki in 1997 stated that, "Education serves society in a variety of ways. The goal of education is to make people wiser, more knowledgeable, better informed, ethical, responsible, critical and capable of continuing to learn" (UNESCO 1997) (http://portal.unesco.org/education/en/ev.php URL_ID=23929&URL_DO=DO_TOPIC&URL_SECTION=201.html. Accessed on March 29, 2012.).

"Environmental Education is an educational process about, for, and through the environment to upgrade the living standards of the people. It aims at improving the quality of life of the people and the quality of the environment with importance placed on Sustainable Development through an understanding and wise management of all the natural resources" (UNESCO 2002: 1–2). Thus, education has to achieve the agenda of sustainability and development as a unified goal, and the agenda for education for sustainable development (ESD) lies in achieving such a synergistic orientation. This means education increases people's knowledge and awareness about social, economic, environmental and associated challenges, develops the necessary skills and expertise to address the challenges, and fosters attitudes, motivations, and commitments to make informed decisions and take responsible action. That is, ESD needs continuous awareness raising and training programmes for achieving sustainable development which can be provided for all ages, at all levels and at all times (UNESCO 1977).

Education must not be equated to schooling or formal education alone. It includes non-formal and informal education modes of instruction and learning, including traditional learning in the home and community as stated in the report of the UNESCO Thessaloniki International Conference (UNESCO 1997). Education has a wider scope than perceived and it is ESD that offers the template to reach out to a wider public in order to address all sorts of environmental and socio-economic issues effectively.

In Bangladesh, the definitions of sustainable development (SD) and education for sustainable development (ESD) have not yet been expressed but as we show in this chapter, the substance and need for it are clearly understood and articulated to some extent in various policies and systems. Generally, people in the country use the Brundtland Commission definition for all purposes. As Filho observed, across countries and international organizations, ESD is one of the most widely used words. In Bangladesh, it is used by technical communities and their associated education systems, in the field of environmental policies and development plans, and extensively in tertiary education courses (Filho et al. 1996; Filho 2000). The existing information, education and mobilization processes also have an important role in the evolution of sustainable development in Bangladesh and the rest of the world

(Romero 1995). Universities and other equivalent institutions play a leading role in promoting environmental ethics and the principles for sustainable development (Delakowitz and Hoffmann 2000). In practice, ESD at the formal, non-formal, and informal levels can play a key role in creating awareness and a motivation to hone people's skills, values, and behaviour for sustainable development.

4.2 Environment, Population and Development

Bangladesh faces a number of environmental problems due to its geographical location and setting, high density of population, poor socio-economic development and inefficient resources management and institutional framework. Bangladesh is primarily a delta. The entire country is inter-connected through river systems and underground aquifers. For a considerable part of the year the land remains saturated with water. This aqueous environment makes it very easy for pollutants to disperse. A National Environmental Management Action Plan (NEMAP) was initiated by the government through the MoEF (Ministry of Environment and Forests) following the commitments made under Agenda 21 at the United Nations Conference on Environment and Development (UNCED) in Rio de Janeiro in 1992. NEMAP identified the key environmental concerns of Bangladesh and provided an action plan to halt or reduce the rate of environmental degradation, improve the natural and man-made environment, conserve habitats and biodiversity, promote sustainable development and improving quality indicators of human life. In the NEMAP environmental problems facing the country were categorized into four broad groups (Huq et al. 1998) (see http://www.eldis.org/fulltext/bcaspop.pdf).

Not Direct Quotation

(a) *Sectorial problems* include natural disaster, water and (Flood Control, Drainage and Irrigation) FCDI, industrial pollution, deforestation, energy crisis, agro-chemicals and land degradation, decline of fisheries resources, loss of biodiversity, health and sanitation, air pollution, urban waste generation, inadequate and poor housing, faulty transport system and lack of environment education and awareness.

(b) *Locational and eco-specific problems* including degradation of wetland, hill cutting, salinity and shrimp cultivation, degradation of coastal and marine resources, char land (a tract of land surrounded by the waters of an ocean, sea, lake, or stream) problem, degradation of upland resources.

(c) *Long-term issues* include climate change and sea level rise, urbanization, regional water sharing and lack of research and development on the issues.

(d) *Institutional issues* include poor institutional setting, lack of inter-sectorial coordination, top-down approach, inadequate local level institution and lack of peoples' participation. There has been a lack of institutional mechanism to deal with inter-sectorial issues at national level.

The country has a very low land to man ratio, with 1142 people per square kilometre; it is one of the most densely populated countries in the world. This population density is three times higher than that of India and nearly five times that of Pakistan (Islam 1999). With this poor land-man ratio, the country is further threatened by both natural hazards and resource over-exploitation. The vast majority of the population lives below the poverty line and almost exclusively off the natural resource base that is also under serious threat (Rahman et al. 2002). A rapid rate of population growth provides a unique setting for examining population-environment linkages. Any increase in population increases demands for consumption of material goods and services that leads to over-exploitation of natural resources such as water, fisheries and forests, leading to degradation of the resource base and the depletion of both ecosystems and the environment of the country. The high and increasing population density of Bangladesh is itself a direct cause of environmental degradation further becoming the cause for recurring natural disasters and a dwindling natural resource base. In the pursuit of development for an improved standard of living and quality of life, environment on which the poor actually depend for their livelihood and wellbeing has been disregarded. Environmental costs are now exerting much greater pressures on natural resources and, especially on the poorer and vulnerable communities of the country. This situation further can be attributed to the poor physical and mental health of the inhabitants which reduces the productivity of individuals (UNDP 1995).

Widespread poverty and illiteracy among the majority of the population increases the country's susceptibility to environmental damage. Particularly, high levels of illiteracy aggravate the susceptibly, because illiteracy reduces the communication of complex information and acts as a barrier for citizens to accumulate knowledge. This further explains the absence of popular environmental protest, giving parties who pollute or damage the environment virtually an open field. These above mentioned factors result in a low level of environmental awareness and the insignificant level of environmental education available to people through schools and other institutions.

It would, however, be erroneous to put the entire blame for environmental degradation on the people alone, because much of it has resulted from such factors as poor markets and policy failures. Given the complex interplay of population, environment and development, the big question is, are there ways to improve the quality of human life while maintaining the ecosystem and environment? Though there is no single solution applicable to all the concerns, there are experiences and lessons from within the country and the outside world to assist in finding solutions for many of the problems in an integrated way and as a comprehensive strategy. Strategies should include and improve the situation of maternal and reproductive healthcare, reduce infant and child mortality, empower women, up-grade the nutritional status of the people, improve sanitation, support education and environmental awareness among the people (Huq et al. 1998: 12–13). All these areas are commonly perceived under environmental education, especially ESD, and are seen as pre-requisites for prosperous economic development. There is an urgent need to develop environmental expertise capable of research and implementation of community education through effective ESD initiatives and programmes. Education and awareness, offer the most, cost-effective and long term solutions to the problems of poverty and population growth and such initiatives and programmes must include support through government, non-government agencies, academic institutions and community/social group networks (Huq et al. 1998).

4.3 Policies Promoting Education for Sustainable Development

The Bangladesh government has several policy statements which have taken cognizance of the significance of education and public awareness in issues related to environment and development. Education has been seen as an enabler in environmental conservation and sustainable development, but there has never been a specific government policy for environmental education and ESD or any guidelines for the environmental professions in Bangladesh. Several laws, regulations and policies have been adapted since the Stockholm mandate (UN Conference on Human Environment) 1987. Even before this date, the first Water Pollution Control Ordinance was brought out in 1973 followed by the promulgation of the Environment Pollution Control Ordinance in 1977 as a part of the evolutionary process of protecting the global environment, natural resources and environment in an integrated fashion. This was much in tandem with the world realizing within 10 years of Stockholm that "…treating environmental concerns in isolation of development needs, such as grinding poverty for a large segment of humanity, was not going to serve either the environment or people" (UNESCO 2005: 26). The idea of environmental protection through comprehensive national efforts was first recognized and declared with the adoption of the Environmental Policy 1992. It covered all geographical regions of the country and fifteen development sectors including agriculture, industry, health and sanitation, energy and fuel, water development, flood control and irrigation, land, forest, wildlife and bio-diversity, fisheries and livestock, food, coastal and marine environment, transport and communication, housing and urbanization, population, education and public awareness, science, technology and research, legal framework and institutional arrangements (Salahuddin 2010).

Furthermore, the Environment Policy, 1992 incorporates the following which come under the purview of ESD:

- "Eradicate illiteracy and create widespread mass awareness regarding protection of the environment and utilization of all national resources in a sustainable and environmentally sound manner.
- Ensure inclusion and dissemination of environmental knowledge and information in the formal and non-formal systems of education and media.
- Encourage spontaneous and active participation of people in all environmental activities.

- Incorporate environmental issues in all training programmes for public and private sector officials and employees including industrial and commercial workers.
- Encourage necessary research and evolve technology so as to ensure long term, sustainable and environmentally sound utilization of all resources.
- Ensure that environmental issues get due consideration in all research activities by research and development institutions" (Environmental Education and Environmental Management in Bangladesh.http://forum.daffodilvarsity.edu.bd/index.php?topic=1086.0 Accessed on March 29, 2012).

This policy reflects the country's acknowledgement of the need for a comprehensive strategic approach to address the environmental challenges and issues of Bangladesh and education and public awareness being one of the strategic areas. This may be hailed as the first policy where environment, development and education feature together, reflecting the pursuit of sustainable development.

The Government of Bangladesh adopted a number of supplementary policies through which environment and development issues are being addressed. These policies have taken cognizance of the concept of sustainable development in principle by recognizing that it is important to balance economic growth, social development, and environmental integrity as these three aspects are irrevocably linked to each other.

Salahuddin (2010: 8) argued that, "The formulated Environment Policy although fairly rich in content is not supported by necessary actions of implementation". According to the Constitution of Bangladesh, some of these policies are not judicially enforceable, however they are still potentially important in guiding and influencing the activities of the concerned Ministries and other governmental agencies.

The Fourth Five Year Plan for Bangladesh (1996) states that ESD would be imparted to the teachers and students at all levels of education and specific measures must be undertaken to ensure participation of women at every level of education. The National Education Policy 2010 (Ministry of Education, Government of the People's Republic of Bangladesh) refers to environmental education and climate change education at several places in a scattered way but nothing specific on ESD. For example Goal 18 of the policy states: "to build students as skilled human resources to fight the challenges of the world threatened by Climate Change and other natural disasters and to create in them a social awareness about environment" (p. 23). All of these are concentrated only on higher education systems, not at primary and secondary education systems (for children below Year 12).

4.4 Environmental Education and Education for Sustainable Development in the Formal Education System

At the beginning of this century, popular environmental movements heightened the awareness of environmental issues and the need for education and skilled environmental management in Bangladesh. Martin has explained how "internal education" is a "vital first step" for any movement: "At the very least, a core of activists must become knowledgeable about the issues so that the case can be presented and argued to the wider public…" (Martin 1988: 202–223). Such movements were popular and involved many sectors of society including academics, professional groups, the general public, many non-resident Bangladeshis, government departments and non-governmental organizations, including, for example, Bangladesh Poribesh Andolon (BAPA) and the Bangladesh Environment Campaign. Since 1995, these movements have provided a common platform for pro-environmental forces in the country. To date, this pressure has been very successful in drawing national attention to a range of environmental issues.

Everett (1972) focused on the role of formal education in environmental movements and has analyzed the potential role of formal education and observes that formal education may be used to institutionalize environmental movements. Extending this observation, formal education could be used to institutionalize the environmental and sustainability causes along with the approaches of the movement itself. Formal education in Bangladesh is a mandate of the government and as the mainstream education it is, perhaps, a ready-made avenue to introduce education related to environment and sustainability. Formal education provides the scope for structured and organized learning. Some researchers argue that formal environmental education helps students to develop more favourable attitudes towards the environment (Bradley et al. 1999; Ramsey and Rickson 1976 as cited in Sarkar 2011).

"The education system consists of general education, technical education and madrasah (religious school) education.General education is the mainstream education in Bangladesh. There are four main levels in the general educational system: primary, secondary, higher secondary/college and tertiary. Primary level schools are divided into: (1) government primary schools, (2) registered non-government primary schools, (3) non-registered non-government primary schools, (4) ibtedayee madrasah (religious schools for primary age children), (5) and kindergarten schools. At the secondary and higher secondary levels, schools and colleges fall into two categories: government schools and registered non-government schools. At the tertiary level, there are general and specialized universities and institutions of higher learning run by the public and private sectors in the fields of engineering and technology, agriculture, medicine and others"(Sarkar 2001: 104).

Sarkar (2001: 104) stated that, "The history of Environmental Education (EE) unfolded following the Qudrat-i-Khuda Commission Report of 1974, which led to the introduction of new syllabi in 1978". Although there is no particular policy document on environmental education in Bangladesh (Salequzzaman and Stocker 2001) in the realm of formal education, the concept of environmental education started in 1978 as a separate subject called environmental studies (Poribesh Shiksha) first in primary stages. In the syllabus of 1978, environmental studies was introduced as a subject at the primary and secondary levels, replacing science and social studies.

"The concept of EE began to take concrete shape during the mid-1980s. In 1986 the National Curriculum and Textbook Board, with assistance from UNDP and UNICEF, improved school curricula, developed teaching aids and materials and trained in-service teachers through a curriculum dissemination programme…The programme was implemented at the primary level from 1992 to 1996. EE is provided in both formal and non-formal schools. All government schools as well as registered non-government schools follow the national curriculum. NGOs implementing governmental projects and programmes in the educational sector also follow the national curriculum. "The Bangladesh Rural Advancement Committee (BRAC) and some other NGOs follow curricula that are different from that of the government" (Sarkar 2001: 106–107). He commented on how a study of school curricula and textual material reveals Bangladesh's EE strategy — infusion, integration, and as a separate course. At lower levels (Grades I and II), infusion is practised. Integration is used in Grades III to VIII. It is taught as a separate course at tertiary education levels. Sarkar (2011) in Secondary Students' Environmental Attitudes: The Case of Environmental Education in Bangladesh reported that environmental education was introduced at the primary level in Grade III by the National Curriculum and Textbook Board and that after the primary level, environmental education is provided to students through a multidisciplinary approach. "At this level, two units deal with environmental education: "Introduction to Environment: Science I" and "Introduction to Environment: Social Science II" (http://www.hrmars.com/admin/pics/95.pdf. Accessed on May 24, 2012).

The main principles for framing the environmental education (studies) syllabus for primary schools are (Salequzzaman and Davis 2003: 4):

- "The pupils will be able to observe and know their immediate environment and develop a scientific attitude in solving their everyday problems;
- Study of separate subjects like biology, physics, chemistry, geography, geology; and social studies; does not help young pupils understand the wholeness of the environment, so an integrated subject is very important; and
- Knowledge about proper use of environmental resources and of their conservation is essential for the maintenance of human life and civilization."

Regarding secondary level, Sarkar (2011) quoted National Curriculum and Textbook Board's Curriculum and syllabus: Secondary level (grades IX-X), "Similarly, at the secondary level, environmental education is provided to students through different subjects, such as language, social science, general science, and biology. These subjects deal with various themes relating to the environment, even though no general objectives of secondary education explicitly state any direct emphasis on environmental education" (http://www.hrmars.com/admin/pics/95.pdf. Accessed on May 24, 2012). Although the environment receives substantial coverage in primary education, the coverage at lower secondary school (Grades VI-VIII) is rather sketchy (Salequzzaman and Davis 2003).

Environmental studies have been introduced in higher education after a couple of decades from the inception of environmental studies at the primary level. However many governmental and non-governmental organizations, semi-autonomous and autonomous institutions were playing a role in providing environmental education (Sarkar 2011) (http://www.hrmars.com/admin/pics/95.pdf. Accessed on May 24, 2012). The formal education sector is often influenced by a political agenda. Governments are obliged to acknowledge strong global developments and display their stance by showcasing the changes introduced in curriculum. For most developing countries, EE and ESD are mere curricular inclusions carried out more or less as an obligatory exercise while receiving development assistance.

Again, ESD, which covers the environmental, economic and social aspects of development, is a recent phenomenon in higher education (tertiary education) where a few universities offer environmental sciences and allied courses. In these courses attempts are being made to integrate the principles, values, and practices of sustainable development into relevant curricular areas.

The Ministry of Education claims that Bangladesh has achieved substantial progress in sustainable development through different educational programmes such as: environmental education (including disaster management and climate change education), non-formal education, vocational training programmes, female education programmes, adult literacy programmes, and inclusive education programmes (Saifullah 2008).

With regard to formal education — primary, secondary and higher secondary levels — as well as teacher training programmes, it can be concluded that, though many issues relating to economic, social, environmental and cultural aspects of sustainable development are incorporated in to the curriculum as specific topics of study, these have not been truly integrated as crosscutting themes.

Saifullah has identified a large number of specific topics in the curriculum that relate to ESD, including poverty alleviation, rural development, sustainable urbanization, human rights, human security, healthcare, good governance, environmental protection, natural resources management, climate change, disaster prevention and mitigation, biodiversity, natural heritage, peace, tolerance and respect towards others' culture and religion, cultural heritage (tangible and intangible) and tourism (Saifullah 2008: 56–60).

Van Hemert et al. (1995) as cited in Saeed et al. (1998: 43–56) commented on the common perception of education which is a seen as a one-way flow of information, usually in educational institutions, especially schools, where ESD can be a two-way communication with full participation and learning by people of all ages. The educational process itself becomes sustainable when the participants take responsibility and lead the process themselves i.e. a two-way process. Thus, ESD should not be confined to schools, but is an important tool for managers, civil servants, community groups and NGOs alike, enabling them to implement policies to protect the environment.

4.5 ESD Through Non-formal and Informal Approaches

Non-formal education is defined as, "… any organized, systematic, educational activity carried on outside the framework of the formal system to provide selected types of learning to particular subgroups in the population, adults as well as children (Coombs and Ahmed 1974: 8). It is based strongly on the pedagogy of transformative learning and empowerment which is central to ESD.

The formal school system in Bangladesh still has an inadequate delivery of education and is largely geared to develop human resources for economic growth. All aspects of formalization within the system and widening the scope and increasing the depth in terms of quantity and quality need steering of the entire system which in itself is a colossal task. It is due to such system-linked inertia that formal education systems will take time to espouse ESD in any real sense.

For a country like Bangladesh, local sustainability goals can best be achieved through formal, non-formal, and informal means of education. However, the fact remains that the entire country has to be involved in the process of sustainable development. Non-formal education provides that second best chance. Formal education has as set of problems related to inadequate infrastructure, curricular issues in terms of relevance, lack of teacher preparation, accountability, absenteeism, learning only based on textbooks (Shohel and Howes 2006). Internal efficiency of the school system is low, reflecting high dropout and repetition rates. "A child joining primary school at the age of six at Grade I should normally complete primary education by age ten. But in reality, it takes, on average 6 years for a child to reach Grade IV (by the time he/she is already 12 years old) and up to 8.7 years to complete the 5-year cycle (by the age of 14+)" (Ministry of Primary and Mass Education 2003).

Other traditional barriers such as poverty and inequity act as strong reasons for pulling down enrollment figures. The inadequacy of formal education systems and other socio-economic factors tend to keep a large proportion of people away from learning opportunities.

Non-governmental organizations (NGOs) in Bangladesh are encouraged to undertake a wide range of environmental action programmes. Shohel and Howes (2006) stated in a paper Nonformal education for sustainable development: A Bangladeshi perspective mentioned that about 450 NGOs run educational programmes in the country (http://www.unescobkk.org/education/apeid/apeid-international-conference/10-th-apeid-international-conference/. Accessed on March 29, 2012). In addition to these NGOs, community service organizations such as Environmental Awareness Club, Earth Club, Rotary, Lions, and Apex, are active both in environmental improvement projects and environmental awareness programmes.

Since the 1960s, non-formal education (NFE) has comprised a wide spectrum of educational and training activities organized outside the formal school system (Shohel and Howes 2006). In this non-formal sector, education focuses on the learners' needs, stresses outcomes, and in many situations uses the learner as a resource

(Peace Corps Centre for Field Assistance 2004). Many non-formal education programmes seek to integrate traditional environmental wisdom with insights and skills available from the sciences. Innovative learning methods in non-formal schools are aimed at the development of practical skills, including matters of health, sanitation and literacy, to be applied in real life situations. Freire (p. 9–10) used "problem-posing" to raise awareness of social issues and to stimulate action. Using a process of problem analysis, reflection, and action, his approach to education was based on the belief that community members need to be encouraged to think critically about problems in their daily lives in order to make decisions and take action (Peace Corps Centre for Field Assistance 2004). For instance, participants in NGO-based environmental education are assigned to take the lead in installing sanitary latrines, mobilizing children to be vaccinated against deadly childhood diseases, making provision for safe water, motivating villagers to make compost, and providing nutrition advice to mothers. In addition, people of the community have been mobilized for energy-efficient and safe household activities such as the use of improved traditional earthen cookers (Siddiqui 2003). A strong theme for non-formal education programmes run by NGOs in Bangladesh is education itself – addressing illiteracy, contributing to the basic education of children, especially those from the poorest families, promoting the participation of girls in education; empowering women; and supporting the government's universal primary education programme (Shohel and Howes 2006) (http://www.unescobkk.org/education/apeid/apeid-international-conference/10-th-apeid-international-conference/. Accessed on March 29, 2012).

Both NFE and informal learning are approaches for raising environmental awareness. Informally, every day, we learn. Lifelong learning is usually unintended and the content is determined by the individual learner to suit his or her needs as they arise. These have a long tradition in Bangladesh, particularly through folk and religious traditions. Local methods of fishing, farming, health care and other technologies have developed over long periods in various areas of the country. This traditional wisdom and these technologies were often based on environmentally sound practices, such as integrated organic agriculture that is now popularized as 'permaculture' in the West.

The informal spaces and time provide a continuum for learning that is directed to address environmental, social, cultural and economic issues. For example, by tradition, rural people do not eat catfish during the Bengali month of Chaitra (mid-March to mid-April). This is the period when the water bodies are relatively dry and the preservation of such fish for the next season requires that their stock be not exhausted during this season. This practice has relevance in industrial times due to the fact that during dry periods, there is a relative increase in levels of pollution.

Versified (prose available in metrical form) wisdom, such as the Khonar Bachan (Nawaz 1989) has for centuries guided agricultural practices in Bangladesh. Khona is a mythological wise lady. According to some authorities, Khona is a historical person, the wife of the astronomer Mihira (a court scholar to Vikrmaditya, 380–413 AD). Khona was herself an accomplished mathematician and astronomer. Khona has been considered by other philosophers to be an imaginary person, her sayings symbolizing folk wisdom based on centuries of farming practices (Nawaz 1989).

Other folk sayings relate to health care, protection of crops and conservation of resources. Local plants like Neem *(Azadirachta indica)* and Bishkatali *(Polygonum sp.)* are used for protecting food grains from pests during storage in many areas of Bangladesh.

From the above discussion, NFE and informal education are an extremely necessary means of sustainable education for Bangladesh and find a close resonance with essential qualities of ESD as compared to formal education systems that have to be externally driven to be reoriented to espouse ESD (Shohel and Howes 2006)(See http://www.unescobkk.org/education/apeid/apeid-international-conference/10-th-apeid-international-conference/. Accessed on March 29, 2012).

4.6 Climate Change (CC) and Education for Sustainable Development

Over recent decades, scientific research has demonstrated how climatic changes have important impacts on the livelihoods of people around the world. For most developing countries their levels of structural and social vulnerability are a dangerous combination and a formula for impacts of higher magnitude (Carvajal 2007). The adverse effects of such forces as the increase in global temperature, change of climate and sea level rise are becoming evident in Bangladesh.

Bangladesh is situated at the unique juxtaposition of the composite, sprawling, interlinked Ganges-Brahmaputra-Meghna river systems, the second largest river system in the world, which drains an area of 1,086,000 square kilometres from China, Nepal, India and Bangladesh. While all countries of the world – whether developed, developing, or least developed – are susceptible to climate change, Bangladesh is highly vulnerable to climatic manifestations that include short-term and long-term impacts of climate change. The reasons lie in its unique geographical location, hydro-geological characteristics like dominance of floodplains, low elevation from the sea and the socio-economical characteristics, including high population density, high levels of poverty, and overwhelming dependence on the natural environment (Salahuddin 2010). Bangladesh could, perhaps, be considered an exemplar case depicting how the unique geophysical location endowed it with rich biological diversity that has been a source of livelihood and sustenance to its people. Now the same resources, under the impact and threat from climate change, are eroding people's economic security and ecological wellbeing. The country lends itself to a study to understand the many complex interactions between climate change, population, and environment. Climate change will intensify many existing environmental, social, and economic problems and natural hazards bringing these to the tipping point. At one level, the interrelated and interdependent environmental, social, and economic factors are themselves recognizable barriers to growth and development and these, on being aggravated by climate change, are turning out to be undisputed barriers to qualitative and holistic development or sustainable development.

Though circuitous, it has been recognized in the *Poverty Reduction Strategy Paper* (PRSP), a background paper which informed the drawing up the National Adaptation Programme of Action (2005). It drew attention to the importance of human resources development as a coping mechanism for combating climate change and an enhanced investment in education as the means to achieve better human resources. The PRSP recognized the direct links between poverty and vulnerability to natural hazards. It also proposed a comprehensive and anticipatory approach to reduce Bangladesh's vulnerability: "… to reduce vulnerability to natural, environmental and human induced hazards through community empowerment and integration of sustainable risk management initiatives in all development programmes and projects" (www.thegef.org/…/GEFProjectDocuments/…/12-10-2008%20ID3873%2… Accessed on March 29, 2012).

The National Adaptation Programme of Action (NAPA) (Ministry of Environment and Forest 2005) identified two major types of strategies to deal with climate induced variability and extreme events: intervention strategies and facilitating strategies. Inclusion of climate change issues in the curriculum at secondary and tertiary educational institution was seen as a facilitating strategy underpinning the imperative that the younger and future generations are made aware of the problem (and solutions) of the climate change impacts. The Board of Education and universities were identified as the implementing agencies. This involved developing an appropriate curriculum on climate change impacts and adaptation for primary school students and for secondary school students and then incorporating these into the school curriculum.

Bangladesh is a signatory to the United Nations Framework Convention on Climate Change (UNFCCC) and the Kyoto Protocol, and the country has incorporated science and technology of climate change mitigation and adaptation into the educational curricula of universities and agencies such as concerned ministries, departments, and organizations for smooth implementation (UNFCCC 1998).

The NAPA submitted to UNFCCC identifies fifteen areas where necessary programmes will be undertaken. Essentially, it recognizes that trade-offs among options at the level of the whole society, and a cooperative approach involving each and every citizen and nation-state of the world is needed. Further, citizens have to be well informed about the risks involved through appropriate education and awareness programmes to generate skills and societal support for a sustainable solution. This education ultimately tends to ESD.

The COP (Conference of the Parties) eight of the UNFCCC in 2002 adopted a 5-year New Delhi Work Programme; again an Amended New Delhi Work Programme (WP) was adopted at the 13th COP at Bali in December 2007. The WP contains six inter-related elements: education (youth in particular is the target); training (scientific, technical and management personnel); public awareness; public access to information (CC:iNet--*Climate Change Information Network of the UNFCCC*); public participation; and international cooperation. Implementation of these conditions is now going on at different universities, organizations and departments in Bangladesh. CC:iNet contains a wealth of information on ways parties are improving the awareness and participation of the public in climate change matters worldwide. It is intended to strengthen efforts by Parties to the UNFCCC to implement activities under Article 6 of this Convention, which deals with 'Education, Training, Public Awareness, Public Participation, Access to Information and International Cooperation'. All of these activities are also a part of ESD initiatives and programmes in Bangladesh.

4.7 Community-Based ESD

Sustainability as a holistic concept demands that all strategies for sustainable development be focused upon and utilized. Communities offer themselves as viable units for planning and executing interventions in common socio-economic and ecological aspects through organizational frameworks. "The term 'community' is used here in a broad sense to indicate those who share a common milieu and therefore face an interlocking set of challenges in sustainable development" (UNESCO 2005: p. 11).

It is well known that Community-Based Development (CBD) and its more recent variant Community Driven Development (CDD), are among the fastest growing mechanisms for channeling development assistance (World Bank 2003). The doubts expressed about the faith in communities being able to address the overarching issue of sustainability, for instance, in restoring population-resource imbalances, coping with disasters, mitigating effects of pollution and so on, due to the fact that communities may not be able to achieve consensus, scale, and control over externalities, to name a few.

Sen, cited in World Bank (2003), stated the influential effort to shift the focus of development from material well-being to a broad based 'capability' approach. This shift has deeply influenced many in the mainstream development community. This paradigmatic shift is an enabler for sustainable development and is the cornerstone of ESD. Put in simple terms, community empowerment and lifelong learning can make communities work cohesively and play an active role in addressing issues that usually act as barriers to sustainable development. Another important aspect of ESD is 'participation' and it was Robert Chambers who led the inclusion of 'participation' as a crucial aspect of empowerment (World Bank 2003).

In summary, community-based strategies can play a critical role by bringing many socio-economic and environmental issues to scale, and provide a common organizational framework wherein capacity building and skill development have a well-defined purpose, easy to acquire and used for collective action.

The Grameen Bank project of Mohammed Yunus is a point in case. The Grameen Bank (Grameen means rural) of Bangladesh, started by Professor Muhammad Yunus, is a microfinance organization and community development bank which provides small loans. The system of this bank is based on the idea that the poor have skills that are under-utilized. A group-based credit approach is applied which utilizes the peer-pressure within the group to ensure that the borrowers follow through and use caution in conducting their financial affairs with strict discipline, ensuring repayment eventually and allowing the borrowers to develop good credit standing. In October 1983, the Grameen Bank Project was transformed into an independent bank by government legislation (The organization and its founder, Muhammad Yunus, were jointly awarded the Nobel Peace Prize in 2006).

The Grameen Bank has introduced a mechanism under which credit can be provided to the poorest of the poor on a group liability basis instead of under individual collateral. Although, from the point of view of profits, the Grameen Bank is not yet a viable institution, empirical evidence suggests that the Bank's credit programme has significantly improved the socioeconomic conditions of its borrowers.

The International Implementation scheme of DESD recorded that "Ultimately the DESD aims for ESD to be implemented in thousands of local situations. This will involve the integration of ESD into a multitude of different learning situations". It urges the need to build upon, "the learning from years of environmental, health, peace, economic, human rights, and development education networks around the world that for many years have used innovation to deliver valuable services in difficult situations" (UNESCO 2005: 20).

4.8 ESD Through Broad Based Environmental Movements

Environment and sustainable development needs to enter into the collective consciousness of the people and government of Bangladesh in their pursuit of economic growth. Education and learning are keys to this. While learning in the formal education sector offers a receptive structure to introduce ESD, informal learning, being contextual, has its potential for inculcating values and sensitivities needed for ESD. Environmental education and ESD have to enter the agenda of all sections of Bangladesh's civil society. The professional and trade organizations have to take up ESD issues. Literary and cultural organizations are reflections of the society they represent and can influence people by showcasing the challenges and barriers to sustainable development. Educational institutions and students have to assume a leading role by shaping the thought of future citizens. Women and children's organizations have a special role to play. Journalists and other members of the media can be very effective in spreading awareness and mobilizing citizen's action about environmental issues.

The media, particularly the print media are bringing to light many environmental problems. Many civic voluntary organizations are emerging with environmental protection as the goal. There are many ongoing movements focused on specific environmental aspects, such as the protection of Dhaka's greenery and lakes. There are environmental movements outside of Dhaka, in various districts too. As these are precisely the qualities that Bangladeshi people want to see in their community organizations, this nascent environmental movement has a significance that goes beyond ESD concerns, and may be conducive to a process of social regeneration. Such a process will ultimately lead the Bangladeshi political parties to mainstream environment and sustainable development into its mandate. ESD is a lifelong process and it is through such associations, groups and spaces that ESD can get in touch with people.

4.9 Conclusion

This chapter has highlighted how Bangladesh lends itself to be seen as a unique country which needs to espouse sustainable development not as a plan for the future but as a direction to act today—with government, community and individuals being part of this. Developmental challenges like population growth, poverty, livelihoods, climate change and fragile ecosystems have actually widened the narrow mandate of education from just preparing people for economic prosperity.

There is a clear understanding in the Ministry of Education that sustainable development is dependent on human resource development which in turn is dependent on education and training. Formal education, further, is dependent on teachers, curricula, infrastructure and a sustainable environment. The Ministry of Education and other relevant ministries need to rely on the strengths of social, cultural, political and administrative systems for qualitative improvements in education such that it can foster a consciousness for learning for sustainable development. Notably, approaches of non-formal, informal, and community-based education have evolved to be a necessary part of a sustainable education system in the context of Bangladesh. Though formal, non-formal and informal streams of education provide a range of tried and tested means to influence a large segment of the population, their mere use is not proof enough to claim that ESD or EE is happening, particularly with the quality and focus required. Under such circumstances, the challenge is to channel all possible means of education to find a direction towards sustainability, in the larger sense of the word.

References

Atkisson, A. (1999). *Believing Cassandra: An optimist looks at a pessimist's world.* US Publisher: Chelsea Green, and Australian and New Zealand Publisher: Scribe Publications, Melbourne, Australia. http://www.atkisson.com. Accessed 21 Dec 1999.

Bradley, J. C., Waliczek, T. M., & Zajicek, J. M. (1999). Relationship between environmental knowledge and environmental attitude of high school students. *The Journal of Environmental Education, 30*(3), 17–21.

Carvajal, L. (2007). *Impacts of climate change on human development* (Background paper to UNDP 2007, fighting climate change: Human solidarity in a divided world, human development report 2007/2008). New York: Oxford University Press.

CEE. (2009). *Towards a new development paradigm: Education for sustainable development* (India report to the world conference on ESD, Bonn, Germany 2009). Ahmedabad: Centre for Environment Education (CEE).

Coombs, P. H., & Ahmed, M. (1974). *Attacking rural poverty: How non-formal education can help*. Baltimore: John Hopkins University Press.

Delakowitz, B., & Hoffmann, A. (2000). The Hochschule Zittau/Go¨rlitz: Germany's first Registered environmental management (EMAS) at an institution of higher education. *International Journal of Sustainability in Higher Education, 1*(1). http://www.emeraldinsight.com/journals.htm?articleid=1502503. Accessed 1 Apr 2012.

Environmental Education and Environmental Management in Bangladesh. http://forum.daffodil-varsity.edu.bd/index.php?topic=1086.0. Accessed 29 Mar 2012.

Everett, M. (1972). The role of formal education in environmental movements. *Journal of Economic Issues*. Association of Evolutionary Economics. http://www.jstor.org/discover/10.2307/4224145?uid=3738256&uid=2129&uid=2&uid=70&u id=4&sid=21100826100951. Accessed 29 May 2012.

Filho, W. L. (2000). Dealing with misconceptions on the concept of sustainability. *International Journal of Sustainability in Higher Education, 1*(1). Hamburg: Technical University Hamburg-Harburg Technology (TuTech).

Filho, W. L., MacDermott, F., & Padgham, J (Eds). (1996). *Implementing sustainable development at university level* CRE-COPERNICUS. European Research and Training Centre on Environmental Education, UK: University of Bradford.

Heij, E. (2001). *OECD Policy Brief 2001 – Sustainable development: Critical issues*. CSIRO Sustainability Network, CSIRO, Australia. www.sustainability.com. Accessed 23 Dec 2001.

Huq, S., Rahman, A. A., & Mallick, D. (1998). *Population and environment in Bangladesh*. Paper presented at the workshop on Population and Environment in Bangladesh Organized by The IUCN Bangladesh. Dhaka, Bangladesh: Bangladesh Centre for Advanced Studies (BCAS).

IUCN Bangladesh held in November 1998, Bangladesh Centre for Advanced Studies (BCAS). http://www.eldis.org/fulltext/bcaspop.pdf. Accessed 29 Mar 2012.

Islam, A. (1999). *Population, development and environment: The emerging issues. Dhaka: Centre for policy dialogue*. http://www.cpd.org.bd/pub_attach/unfpa2.pdf Accessed 30 May 2012.

Jacobs, P., Gardner, J., & Munro, D. (1987). Sustainable and equitable development: An emerging paradigm. In P. Jacobs & D. A. Munro (Eds.), *Conservation with equity: Strategies for sustainable development*. Cambridge: International Union for Conservation of Nature and Natural Resources.

Martin, B. (1988). *Education and the environmental movement*. http://www.bmartin.cc/pubs/88Lovett.html. Accessed 24 May 2012.

Ministry of Environment and Forest. (2005). *National adaptation programme of action (NAPA), final report*. Dhaka, Bangladesh: Government of the People's Republic of Bangladesh.

Ministry of Primary and Mass Education. (2003). *Education for all: National plan of action II 2003–2015*. http://planipolis.iiep.unesco.org/upload/Bangladesh/Bangladesh%20NPA%20EFA.pdf. Accessed 29 Mar 2012.

Nawaz, A. (1989). *Khonar Bachan: Krishi-O-Bengali Sanskriti (Saying of Khona: Agriculture and Bengali Culture)*. Dhaka: Bangladesh Agriculture Research Council.

Peace Corps Centre for Field Assistance. (2004). *Non-formal education manual*. Washington, DC: Paul. D. Coverdall Peace Corps Headquarters. http://files.peacecorps.gov/multimedia/pdf/library/M0042_nfemanual1.pdf. 24 May 2012.

Rahman, A., Mallick, D. L., Haque, N., & Nishat, A. (2002). Trends in natural resource management in Bangladesh: Looking for integration and a new institutional framework. In *Proceedings of natural resource management workshop; 10 October 2002*. Organized by: Bangladesh Centre for Advanced Studies (BCAS).

Ramsey, C. E., & Rickson, R. E. (1976). Environmental knowledge and attitudes. *The Journal of Environmental Education, 8*(1), 10–18.

Romero, M. J. R. (1995). The role of the university in sustainable development: Challenge and opportunities. *Journal of Higher Education Policy, 8*(4), 26–29.

Saeed, S., Goldstein, W., & Shrestha, R (Eds.). (1998). *Planning environmental communication and education: Lessons from Asia. IUCN-Commission on Education and Communication*. Bangkok, Thailand: IUCN.

Saifullah, Md. (2008). *Environment policy of Bangladesh*. Presentation. In Workshop on South Asian.
Salahuddin, M. (2010). *Environmental policy of Bangladesh: A case study of an ambitious policy with implementation*. Dhaka: University of Dhaka.
Salequzzaman, M., & Davis, J. K. (2003). *Environmental education and environmental management in Bangladesh and their sustainability. Environmental informatics*. Archives, vol. 1 (2003), International Society for Environmental Information Sciences.
Salequzzaman, M., & Stocker, L. (2001). The context and prospects for environmental education and environmental career in Bangladesh. *International Journal of Sustainability in Higher Education, 2*(2), 104–127.
Sarkar, H. S. (2001). Status reports. I. South Asia, (1) Bangladesh. In B. Bhandari & O. Abe (Eds.), *Environmental education in Asia-pacific region*. Kanagawa: IGES.
Sarkar, M. (2011). Secondary student's environmental attitudes: The case of environmental education in Bangladesh. *International Journal of Academic Research in Business and Social Sciences,* August 2011, Vol. 1, Special Issue. http://www.hrmars.com/admin/pics/95.pdf. Accessed 24 May 2012.
Sutton, P. (2001). *Definition of sustainability and sustainable development*. Greenleaf, Australia. http://greeninnovations.asn.au/grenleaf.htm. Accessed 29 Dec 2001.
Shohel, M. M. C., & Howes, A. J. (2006). *Non-formal education for sustainable development: A Bangladeshi perspective*. In The 10th APED international conference: Learning together for tomorrow, education for sustainable development. Bangkok, Thailand. http://www.unescobkk.org/education/apeid/apeid-international-conference/10-th-apeid-international-conference/. Accessed 29 Mar 2012.
Siddiqui, F. M. (2003). *Linking innovation and local uptake in rural development – A case study on energy innovation in Bangladesh*. Ph.D. thesis (unpublished), Institute for Sustainability and Technology Policy (ISTP), Murdoch University, Western Australia.
Sonneborn, C. L. (1998). *Educating engineers about sustainable energy: An overview of the issues*. Occasional paper 3/98, Institute for Sustainability and Technology Policy (ISTP), Murdoch University, Murdoch, Australia.
UNDP. (1995). *Report on human development in Bangladesh*. Dhaka, Bangladesh: UNDP.
UNESCO. (1977). *Intergovernmental conference on Environmental Education*, Final Report, Tbilisi (USSR).
UNESCO. (1997). *Declaration of thessaloniki. Thessaloniki international conference*. http://portal.unesco.org/education/en/ev.php- URL_ID=23929&URL_DO=DO_TOPIC&URL_SECTION=201.html. Accessed on 29 Mar 2012.
UNESCO. (2002). *Education for sustainability, from Rio to Johannesburg: Lessons learnt from a decade of commitment*. Paris: UNESCO.
UNESCO. (2005). *International Implementation Scheme* (IIS). UNESCO: United Nations Decade of Education for Sustainable Development.
UNFCCC. (1998). *Kyoto Protocol: Article 6 and 10e*. UN. http://unfccc.int/essential_background/convention/background/items/1366.php. Accessed on 29 Mar 2012.
Van Hemert, M. et al. (1995). *Reviving links: NGO experiences in environmental education and people's participation in environmental policies. Both ends/SME Milieu adviseurs*. The Netherlands: IUCN
World Bank. (2003). *Evaluating a community based and community driven development: A critical review of evidence*. http://siteresources.worldbank.org/INTECAREGTOPCOMDRIDEV/Resources/DECstudy.pdf. Accessed 29 May 2012.
World Commission on Environment and Development (WCED). (1987). *Our common future*. Oxford: Oxford University Press.
www.thegef.org/…/GEFProjectDocuments/…/12-10-2008%20ID3873%2…. Accessed 29 March 2012.

Chapter 5
Gross National Happiness, Values Education and Schooling for Sustainability in Bhutan

Dasho Karma Ura

5.1 Introduction

Bhutan is a young democracy and its Constitution was proclaimed and a parliament elected as recently as 2008. The Constitution of the Kingdom of Bhutan states, "the State shall strive to promote those conditions that will enable the pursuit of Gross National Happiness" (*Article 9 Principles of State Policy*). From this official pronouncement it is clear that the pursuit of GNH is not optional but mandatory. The Kingdom of Bhutan has been a hereditary monarchy dating back to 1907. Beginning in 1998, it experienced a rapid, peaceful, guided and unflinching transition to democracy solely initiated by the vision and will of King Jigme Singye Wangchuck. The Constitution declares that Bhutan's form of government shall be a "Democratic Constitutional Monarchy," (Constitution of Bhutan Article 1.2) in which "sovereign power [will] belong to the people of Bhutan" (Art.1.1). Within 10 years, in 2008, the country became a constitutional monarchy (Gallenkamp 2010).

The visionary leadership of a Buddhist monarchy played a key role in the country's progress. The Fourth King of Bhutan – His Majesty Jigme Singye Wangchuck – defined and established the concept of GNH. It was during his reign that a road map towards GNH in terms of laws and policies was developed along which Bhutan has continued to travel. GNH was first formulated as a "wholesome idea" in the 1970s (Varma 2010: 13). His Majesty firmly believed that, as a society, Bhutan should have a distinctive, development philosophy if "we are to make our people contented". The simplest message of this idea of GNH is that, "We should not end up by having everything – except happiness and the contentment of a meaningful life" (Ura 2010) (http://www.india-seminar.com/2010/614/614_karma_ura.htm.

D.K. Ura (✉)
President, Centre for Bhutan Studies & GNH Research (CBS), Thimphu, Bhutan
e-mail: dasho.k.ura@gmail.com

R.N. Gorana, P.R. Kanaujia (eds.), *Reorienting Educational Efforts for Sustainable Development*, Schooling for Sustainable Development,
DOI 10.1007/978-94-017-7622-6_5

Accessed on April 15, 2014). He explored a more wholesome approach to guiding governance and development.

His Majesty Jigme Khesar Namgyel Wangchuck, the fifth King, crowned in 2006, has continued to give impetus to GNH. He stated, "Today, GNH has come to mean so many things to so many people but to me it signifies simply – Development with Values. I am confident that the noble goal of Gross National Happiness will be the key to Bhutan's success in maintaining our unity and harmony – indeed our character as a nation". He has also said that "GNH in its simplest form is a balance between tradition and modernity" and the Kingdom follows GNH as an integrated concept of development as opposed to the more widely used metric Gross Domestic Product (GDP). GDP measures economic health in terms of economic activity, it is essentially a production and consumption index of a country (Ura 2010) (http://www.india-seminar.com/2010/614/614_karma_ura.htm. Accessed on April 15, 2014).

5.2 The Concept of Gross National Happiness

The concept of GNH is not new to Bhutan and in its current articulation it has "provided a coherent political basis to the regime" (Mathou 2008: 7). At the core of the GNH concept and its measurement is the view that a more holistic range of human needs should be taken fully into account for a happy and fulfilling life. GNH takes a eudaemonist view of human beings. It stands for the preservation and renewal of a holistic range of capital or resources: ecological, economic, social, human resource, cultural, all of which should be valued and possibly measured. GNH is about making human life more meaningful, fulfilling, and sustainable, through both individual and official actions. At one level, it addresses material needs. On another level it addresses the inner parts of an individual that create a sense of personal well-being.

GNH is not simply about personal happiness. A definition of happiness that fails to take into account the happiness needs of others is irresponsible and self-centred. Happiness blossoms through relationships between people, animals and their environments. It is a state of being that for an individual can be realized partly through the happiness of others. All human beings are central to GNH, just as the recognition of intangible factors is. Those intangible factors often cannot be substituted by income and goods.

The conceptual structure of GNH rests on four main pillars: economic development, environmental preservation, cultural promotion, and good governance. In seeking to define a measurable index of GNH, nine domains and thirty three indicators have been specified. The nine domains are:

- psychological wellbeing;
- balanced time use;
- community vitality;

- cultural diversity and resilience;
- ecological diversity and resilience;
- good governance;
- living standards;
- health; and
- education.

The indicators represent a wide spectrum of human needs associated with being happy in a deep and holistic sense. In some discussions the term enlightenment is used alongside happiness.

5.3 National Happiness: Bhutan's Developmental Philosophy and Goal

The adoption of GNH as Bhutan's philosophy for underpinning development is based on the conviction that, since happiness is the single most important aspiration of every citizen, the primary role of the state must be to create conditions that will enable individuals to transform themselves and society into a blissful state of happiness.

> Happiness has meant different things to different people. For Aristotle, happiness was 'the meaning and the purpose of life, the whole aim and end of human existence.' For His Holiness, the Dalai Lama, 'The very purpose of our life is to seek happiness.' For Mahatma Gandhi, happiness is 'when what you think, what you say, and what you do are in harmony.' According to Hermann Hesse, the Nobel Prize winning German novelist and poet, 'Happiness is a how, not a what; a talent, not an object.' Buddha declared that 'happiness never decreases by being shared.' On the contrary, 'happiness adds and multiplies as we divide it with others.' Not insignificantly, Marie-Henri Beyle, the 19th century French writer better known as Stendhal remarked: 'To describe happiness is to diminish it' (Shiva Kumar 2010: 53).

Happiness has both a non-self-transcendent quality and a self-transcendent quality and the latter has much to do with the good feeling that flows from fulfilling one's social and sociological needs. Material goods are tremendously important to fulfil physical and physiological needs.

Happiness is very closely related to the word *ga* and *ki* in Dzongkha, the Bhutanese national language. *Ga* mean happy and *ki* means peace, conveying the philosophical notion of 'contentment'. Happiness is more than a state of wellbeing it is a state of wellbeing arising out of a notion of contentment – to be content and, therefore, happy. A state of contentment must also ensure that individual happiness does not stand in the way of larger societal wellbeing or public good (UNESCO 2007).

Some social philosophers have held that human beings live in two states; one of doing and the other of being. It is the state of being that humanity relates with welfare and happiness and it is this concern with welfare and happiness that has driven human societies to seek continuously to organize life and work so that happiness and welfare can be achieved (UNESCO 2007).

Shiv Kumar (2010) has pointed to the importance of both external and internal qualities of happiness, referring to the "emotional reactions and cognitive judgments of how life is being led." Happiness measures add a human face to an assessment of development outcomes. Tracking happiness can improve both the performance of government and the practice of democracy. "And we know that more of happiness is almost always better than less of it (which cannot be said of incomes or goods and services)" (Shiv Kumar 2010: 53–54).

Shiv Kumar (2010) has discussed how the State can play a major role in influencing the external quality of life by creating the conditions for happiness. He argues that states can do much to eliminate misery and suffering – and thereby promote happiness. He describes how people who do not get enough to eat, or who fall sick frequently, or feel discriminated against, can never be really happy. He suggests that people feel happy when they enjoy secure lives. Further, dignity and respect make people happy and thus the onus to make public services more responsive lies with the state. To this end, Bhutanese governments, over the years, have dedicated themselves to creating conditions that would enable the citizens to pursue happiness. Thus, GNH is a development paradigm that embraces sustainability and it has a clear purpose and a set of strategies that will enhance and sustain human civilization. Bhutan has become the first country in the world to use a radically new prism to judge the quality of life of a citizen. GNH could be a tool of governance, a vision, a policy guideline, an indicative metric and an alternative development paradigm. For Bhutan it is a guiding philosophy for societal transformation.

5.4 The Emerging Worldview and GNH Worldview

Large scale changes are taking place in Bhutan. One of the major reasons for these changes has been the transformation of Bhutanese citizens into consumers, at the expense of their role as producers, in the information age. Another important contemporary feature is population increase. Globally, this crowding of the earth is a very significant phenomenon that will impact on Bhutan in many unexpected ways. Bhutan's population has been increasing over the last four decades. The risk of climate change will be compounded by the phenomenal rise in population movement in South Asia. The population of Bhutan is likely to stabilise contrary to official estimates, by 2018. Nevertheless, urbanisation will not stabilise unless a conscious, different effort is made in Bhutan's 5 year plans.

Against these changes, the emerging worldview will eventually impinge on Bhutan's values. If the government, the central bureaucracy and the business community – the groups who propel the direction of society – hold a shared worldview, this will be reflected in the policies or major decisions affecting the country. Increasingly, the worldview held by these groups begins to define social and cultural aspects and shapes aspirations that are decoupled from cultures. Together, these three groups are the main source of any new direction; their consensual worldview matters immensely.

How their worldviews can be aligned with GNH is crucial. At an official level, GNH is about removing obstacles of a public nature to collective happiness through policies, programmes and associated public expenditure. Official actions create the conditions either for or against the success of society's striving for contentment. Government decisions and associated public expenditure (currently about 25 % of GDP) are vitally important. A dominant theme in the emerging consensual world-view in Bhutan is that free trade, foreign investment and the free market – all of which lead to a very open economy – are desirable and that official policies must be aligned with this trend. As Sandel (1998) has shown, such a worldview has its origin in the idea that the free market approach is best, and the role of the government is to address market failure, simply by mimicking the market. If a government implements neo-liberal prescriptions the consequence will be that it will be very difficult to claim the identity of Bhutan and GNH since the introduction of a free market will make Bhutan similar to other nations.

5.5 The Implementation of GNH for Achieving 'Wholesome' Development

The focal point of socio-economic development planning in Bhutan, next only to the Cabinet, is the Gross National Happiness Commission that was created by His Majesty King Jigme Khesar in 2007. While conceptualisation and field surveys on GNH have been carried out by the Centre of Bhutan Studies, implementation of GNH through ministries and other agencies is the joint responsibility of the Cabinet and Gross National Happiness Commission. This institutional arrangement underlines the fact that any new goals are to be expressed in terms of policies, and policies must be embedded in new institutions to carry them out. Envisioning a new class of government institutions to reflect the thrust of GNH will become an important part of GNH institutional restructuring if GNH is to gain deeper traction. The present institutional structure, which closely corresponds to and echoes the sectoral composition of GDP in terms of ministries looking after such aspects as agriculture, fisheries, forestry, electricity, mining, manufacturing, and banking with a focus only on material aspects of reality, misses certain relational and intangible factors crucial to happiness. Thus it may not be a pure fantasy to contemplate a new organization such as the Ministry of Psychological Wellbeing and 'Relationality' in the distant future.

GNH is a dominant discourse in Bhutan, with each voice trying to interpret GNH in a way that reflects a particular view of life aspirations – from conscience-concerned lamas, cut-throat businessmen and microphone-loving politicians to procedure-bound bureaucrats and investor-inviting 'globalizers'. The government in 2014 was elected predominantly on a GNH manifesto. Although there is a broad consensus in Bhutan society generally that GNH is a worthwhile aim, new policy directions sometimes provoke subtle disagreements. Such innovative policies as those associated with liberalisation, the market mechanism, free trade and foreign direct investments have been initiated. Debates about GNH are a part of deliberative democracy. To discuss GNH is to discuss the future of Bhutan as a state as well as what human life is for. The debates are about the meaning of change in people's lives, particularly about whether proposed changes would enhance happiness in a holistic, eudaemonist sense. It is of the utmost importance that this important question be posed: What is the relationship between happiness and wellbeing, on the one hand, and economic development on the other? This is the question that lies at the heart of discussions in the global community about schooling for sustainability.

Internationally, less attention has been paid to happiness at the official level. Traditional spheres of government consist of conventional public goods, such as security, economy, social justice, public infrastructure, health and education. Happiness is explicitly not one of them; it is subordinated to the private realm. Public policies and government expenditure priorities are important determinants of the conditions of happiness. So the state ought to consider these conditions of happiness. Left as an individual striving, the chances of policies directed towards attaining happiness succeeding are lower if policies do not take account explicitly of this universal pursuit of individuals.

To have a better system for planning and vetting project proposals, GNH-sensitive project screening tools, differing from one agency to another, have been developed and introduced as an experiment. For example, GNH project screening tools for the health sector are different from those used by the hydro-power sector. However, it will take some years for these particular project appraisal tools to gain wider acceptance in Bhutan, in place of other ways of evaluating the desirability of programmes and projects.

Among the many royal initiatives, the educational reform launched through the Royal Education Council in 2007 is of considerable consequence. In the context of GNH, the weaving of its inherent values into teacher training, school classrooms and textbooks is of vital importance.

Education systems have to pay attention to values, and teachers (*Ston-pa*) have to be involved in instilling these values. Clearly, the curriculum has a direct impact on both teachers and students and it can be negative if they are required to use inadequate textbooks.

What is values education? Values education is not new to the world. It is taught in various countries, but the vocabularies used to refer to it are different. It is known as moral education in Japan, Korea and some European countries and as character education in the USA. Some countries use such terms as civic education, citizenship education and ethics education.

In 2010 the Bhutan Curriculum and Professional Support Division of the Ministry of Education published *Educating for GNH, Refining Our School Education Practices* (Available at: http://www.education.gov.bt/gnh/guidebook/GNH%20Guide%20book.pdf). This publication suggests how GNH values and principles can be infused into the school curriculum as a whole and in various subjects (Namgyel 2010). The simplest idea of values education is the creation of a set of beliefs and attitudes that can be accessed as a person's character and personality unfold. These beliefs and attitudes should influence personal behaviour and actions in a positive manner and direction. In a deeper sense, there is no imposition of values from the outside. Values education should be designed to help students to unfold and develop themselves ethically and realise their ethical potential, despite the many distractions during their early years in Bhutan – and especially in contemporary urban Bhutan. At another level, it is about exploring common values that will help create a positive future for all by first articulating values, and then determining how to apply them to both personal and collective conduct.

According to the foundations of Piaget (1932) and Kohlberg's (1981) heuristics, values education can be made progressive. In Kohlberg's view, values education moves from particular to formal thinking – as children grow up. Values education can therefore be structured from pre-conventional to conventional moral reasoning. In the work of Piaget and Kohlberg a clear prescription of moral development in terms of children's age specification is lacking. If the Buddhist orienting framework is used, moral development involves working on the development of emotional intelligence. Here the focus is on overcoming unwholesome emotions through imagination or mental exercises. The Buddhist approach to moral development places a sharp emphasis on reducing negative emotions to embrace inclusiveness and the individual's relationship with the community. This is a central concern in discussions about schooling for sustainability and education for sustainable development.

So, how is it possible to make positive values not only clear but fundamental beacons of progression in children? This not only involves lessons in values education in schools, but also the values adults transmit to children. These values must be constantly demonstrated through adults' actions, decisions, emotions and behaviour. Parents cannot avoid setting examples. Children and teenagers assimilate a great deal from the world outside the school. Should their parents' values, emotions and behaviour be at variance with what they learn in school, they will discover that adults themselves are in conflict with the values that they are being taught to embrace. The border between the behaviour of adults and children is porous, and adult behaviour is bound to spill over and influence children. As a result, values education undertaken in school is only half the challenge. The other half is undertaken outside the classroom.

5.6 Linking Societal Vision with GNH Values

Values are linked to vision. The vision of the government is GNH. It follows that certain values consistent with this GNH vision have to be inculcated in students. For instance education can play an important role in the new conceptualisation of GNH as observed by McDonald (2005). He stresses how equally and critically important are the interactions between the four interdependent domains – environment, economy, governance, culture – and self. He suggests that these domains could usefully be integrated into the mainstream curriculum to much greater effect.

If educationists want the concept of a 'GNH society' as a goal for their textbooks, then the ideal state and the future that is best for Bhutan have to be clearly outlined. In other words, a 'destinational' vision is needed. In planning for the future, it is very important to understand the kind of world in terms of the valued outcome that the state wants and then plan how to act to proceed towards this goal. There is a need for the state to outline clearly what is wanted for the key elements in society, sector by sector, *dzongkhag* by *dzongkhag*, Five Year Plan by Five Year Plan. It is insufficient to describe contemporary situations in the way that textbooks currently portray them.

In defining a vision for GNH and values education, it is necessary to take into account predictions and theories about the future. Such a vision for Bhutan must be located in a common, global future that embraces and celebrates sustainability, peace and happiness. The current economic system faces considerable challenges in the medium term, over the next 50 years, from such forces as climate change, population increase, resource scarcity, population migration and environmental degradation. It is essential for these issues to be included in school curricula and in school textbooks.

Textbooks that include explicit values education therefore have to address the particular positive values of Bhutan while foregoing those that appear obstructive towards ethical progress. The direction should be from the particular to the universal and not from the universal to the particular. Particular values have to be imparted because these are the foundations that give citizens a cultural identity. GNH has a developmental philosophy and so the values that are contained in it must be reflected in textbooks. There must be correspondence and coherence between these two: values and visions mutually supported by the other. This may draw criticism from those who will contend that, in a democratic country, textbooks should be value neutral. Those who favour such views may go so far as to say that schools should not promote values, not because values are not needed per se, but because they cannot agree about them. This represents, in free market parlance, a 'deregulation of values'. Each value is a priority from an individual or a group's point of view, but the priority given to particular values by individuals and groups is likely to vary. However, since GNH as a value system is accepted in Bhutan though it is only obligatory, not mandatory, something serious should be undertaken at the textbook level.

5.7 GNH and the School Curriculum

One way of thinking about GNH in the school curriculum focuses on the way that the values associated with each of the nine GNH domains, referred to above, should be specified and practised. These nine domains have been specified by the government as constituents of happiness. Happiness, in this context, is perceived as being multi-dimensional. If a person gets various values elements under each of these nine aspects of life right, the chances for happiness will be much higher. In reality, what is most important is the inter-relationship between these domains rather than the domains themselves. This inter-relationship is non-linear. Table 5.1 onGNH Pillars, Domains, Values and Practices demonstrates the inter-relationships between pillars, domains, values and practices. It serves to show how the theory underpinning pillars and domains – the foundations and super-structure of the GNH concept – can be translated into practices commonly found in educational institutions.

Education for sustainable development has been based on the three foundations of sustainable society, sustainable environment and sustainable economy. Much discussion has focused on the latter. To a very large extent this reflects the continuities between educational thinking and practice in the broad field of environmental education. With this in mind, the pillar titled 'Conservation of Environment' is worth highlighting. This pillar is linked to the domain titled 'Ecological Integrity' and the 'Practices' highlight a range of issues that one would expect to find in mainstream social studies, geography and, to some extent, moral education curricula.

There has to be a correspondence in the values inculcated in schools with the values in other sectors of society, such as the economy or industrial activity. If this correspondence is lacking, the relationship will break down. If we get it right in terms of values, but fail to get it right in education again an adverse relationship will develop. Thus, the inter-relationship between these variables that point to the profound interdependencies between various aspects of our life – and the lives of others – is important. The structuring of values according to domains should be viewed merely as a heuristic device: it should not isolate domains into mutually exclusive spheres in practice.

5.8 Values Education Content in a Sample of School Textbooks

In the context of GNH, the weaving of its inherent values into teacher training, classroom teaching and textbooks is of vital importance. At the behest of Gyaltshen Penjor and Tashi Wangyal of the Royal Education Council in 2009, the author was invited to prepare a research proposal focused on values education. This proposal was a beginning of a new focus on the complex task ahead of re-orienting values education. For the education system to pay attention to values is central to a cultural perspective strongly influenced by Buddhism.

Empirical evidence was drawn from a school-focused study conducted by researchers from the Centre for Bhutan Studies (CBS). They reviewed 3800 pages in 27 textbooks commonly used across the social science subjects to obtain a synoptic perspective on what was happening in school classrooms. This study was invaluable in providing an overall impression of the current situation in schools.

Table 5.1 GNH pillars, domains, values and practices (Ura 2009)

Pillars	Domains	Values	Practices
Equitable and sustainable socio-economic development	a. Living standard	Self-reliance, security, quality sustainability, adequacy	Food security, income sufficiency (absolute and relative), financial security, housing
	b. Health	Vitality, fitness, soundness, self-worth, prevention, precaution, non-malignance	Yoga and other physical activities, well-balanced diet, avoidance of intoxicants (drugs, alcohol etc.), avoidance of risky behaviour, public health, mindful consumption
	c. Education	Creativity, openness, diligence, insightfulness, perseverance, patience, creative thinking	Historical literacy, civic literacy, cultural literacy, ecological literacy, food and nutrition literacy, indigenous knowledge literacy
Conservation of environment	a. Ecological integrity	Interdependence, eco-consciousness, sustainability, non-doministic, non-utilitarian, aesthetic, naturalistic, reverence	High air quality, high water quality, conserve biodiversity, retain wildlife species, efficient energy use, proper waste disposal, eco-friendly mass transportation system, eco-friendly industry, eco-friendly foundations of urban sprawl, land use
Preservation and promotion of cultural heritage	a. Cultural diversity and resilience	Identity, dignity, non-alienation	Arts and architecture, media, artisan skills, dialect proficiency, games, tshechus, celebrations
	b. Community Vitality	Altruism, trust, reciprocity, fairness, fidelity, family closeness, solidarity, equality, unity, hospitality, cooperation, honor, sociability and cohesion	Socialization, volunteering, donation, labour exchange, community participation, strong family ties
	c. Time use	Free from stress, serenity, tranquility, bonding, healthy lifestyles	Socialization, work-life balance, recreation and leisure satisfaction, community service
	d. Psychological well-being	Compassion, generosity, forgiveness, calmness, gratitude, taking account of karma, empathy, truthfulness	Meditation for mental training, prayer-ritual as attitude training, pilgrimage as an aid to meditation, assimilation and physical activity, volunteering, donating

How, exactly, textbooks are translated into classroom teaching is difficult to assess; those actually engaged in teaching, obviously, have a clearer understanding.

How can values education be taught? In principle there is both a teacher and a textbook focus. There is a teacher focus, but no textbook focus in some countries, such as South Korea. Only certified teachers are allowed to teach moral education in South Korea. Further research needs to be carried out to know whether there are other countries following a similar approach to this. In Bhutan, there is neither teacher nor textbook specialisation on values education. With regard to values education in Bhutan, teachers have no specialism in this field, they are generalists. They attempt to teach values education if the textbook contains passages related to it. Hence, examining textbooks for their value contents becomes important.

Whether we have formal values education or not, John Dewey (1916) wrote that education is not value neutral. There is a hidden values education, according to him; it is just a question of what kind of values and to what degree values education exists. Dewey claimed that the hidden values curriculum, usually referred to as the hidden curriculum, exists in the systems of symbols, disciplinary codes, hierarchy, ceremonies, and rituals that heighten a pupil's collective identity and values. The atmosphere of any school is thus not neutral. But equally, there is a view that those who design values education also cannot be free from values themselves.

How is values education imparted in Bhutan? In order to carry out values education it is critical for teachers, parents and students to know the value content of the textbooks that are used for teaching. One can argue that the content of textbooks is central, indeed crucial, in influencing both the role of teachers and students in values education, as textbooks are pivotal to both teachers and students. Evidence from the study conducted by CBS shows that the content for imparting values is dispersed and scattered through textbooks.

The twenty-seven textbooks chosen for investigation comprised the language, science and social studies textbooks of classes 4–10. It is important to understand how subjects taught in these classes are structured. Fig. 5.1 summarizes this structure. Two subjects, *Dzongkha*, the national language of Bhutan, and English, are present throughout schooling. Social Studies is taught as a single subject through a single textbook, though of different standards for each class, from classes 4 to 6 in primary schools. In classes 7 and 8 (upper primary), Social Studies is divided into three subjects: Geography, History, and the Earth and its People. Geography and History continue through classes 7–10 (upper primary and secondary schools), but the Earth and its People is once again divided into Economics and Civics. In Science, there is a general textbook for classes from 4 to 8 but from class 8, Science is divided into Physics, Chemistry and Biology.

All of the textbooks in the study have sections incorporating values education. The twenty-seven textbooks were read to glean information on the values education they contained. There were certainly other things associated directly with values being taught which were not reflected in textbooks. So the analysis actually focused only on actual evidence in terms of what was presented in the textbooks. Several conclusions could be drawn from the content through this exercise.

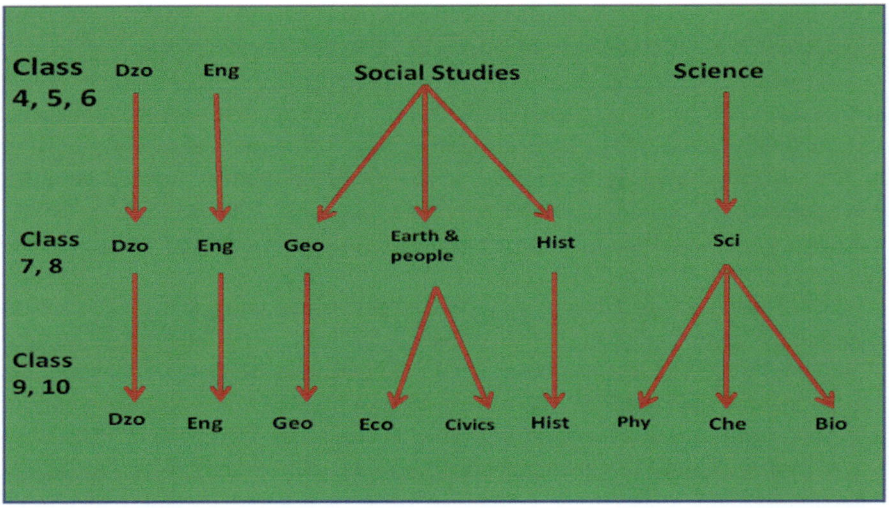

Fig. 5.1 (Ura 2009: 14) (legend--Eng: English, Geo: Geography, Hist: History, Sci: Science, Eco: Economics, Che: Chemistry, Bio: Biology)

The expected areas related to moral and social character are present in the textbooks by being woven indirectly into narratives, stories and biographies. This is the usual approach followed in the design and delivery of values education. The question was: To what degree was values education present? Values such as social conformity and respect for hierarchy were emphasised. In fact, it could be argued that hierarchy was emphasised too much, in the sense that there seemed to be an excessive detail about high post holders such as *dzongda* (district administrators), *thrimpon* (judges), ministers and their ceremonial attire like *kabneys* (scarves). Textbooks have built up the importance of civil servants excessively in the eyes of students. Bureaucrats and politicians are not the only ones who are patriotic and are not the only objects of moral learning. Going by news reports, as well as the perceptions of the people covered in the GNH survey, they are not the most respected members in society, which is in contrast to the textbooks' portrayal. Textbooks should not be the space where politicians and bureaucrats enjoy such prestige in the eyes of the students. It is also unfair to present them as more patriotic or upright than people from other walks of life. Building up the image of these groups, even though desirable, misses the opportunity to explain why such conformity is essential. This explanation is lacking, the emphasis is on description. This criticism of description without reasoning applies to other topics. On the one hand, there is a great deal of recurrent emphasis on festivals and national holidays. On the other hand, there is no explanation of these in relation to values.

With regards to traditions, traditional games and sports appear frequently in textbooks, dispersed in subjects such as geography, history and social studies. Personal hygiene is emphasised in the lower primary school textbooks. Although this is necessary, more complex issues could be taught as students climb up the education ladder. A simple example could be why water is dirty and polluted; it is dirty and undrinkable because someone or some establishment is polluting it upstream. That way we can study various economic and industrial issues to understand why water becomes polluted. But nothing of this sort appeared in any of the textbooks that were reviewed.

Structural issues and causal linkages were avoided. Similarly, in the lessons related to the environment, say, mining as a contributory cause of river pollution is not discussed. There are some sporadic discussions on the harmful impacts of tobacco and alcohol, but again the issue fails to be embedded in the larger idea of why people become addicted. The problem is again not necessarily that of individuals, there is an industry producing alcohol that is also involved. Alcohol consumption is presented as an individual's problem, yet there are bigger causal agencies like distilleries and bars which are directly involved in the abetting and aiding of addiction for commercial reasons. The point is that if there were no production and distribution on a commercial scale, there would be far less consumption. In all coverage of harmful substances, there is no morally engaging discussion that would bring to the surface ethical choices.

Stories based on mythologies and morals, and on Buddhist principles such as compassion and generosity are present, but textbooks do not provide students with any techniques or training in ways of increasing their motivation, not even at class 10 level. As a result, the discussion on values remains inadequate and underdeveloped. So it has been, in a way, just the delivery of jargon.

Mutualism, which is a business-like approach ("You help me, I help you, and often you help me more than I help you"), is covered quite strongly in textbooks in various subjects. Another area – reciprocity and exchange – is particularly emphasised, especially with respect for parents. Respect for the monarchy has to be advocated in a stronger and clearer way, given the centrality of this institution. Parental repayment of kindness is sharply focused, but again it is unlikely to hold the students' attention deeply because there is no grounding of this concept in a moral argument.

The vocabulary used in delivery of values education is important to clarify conceptual explanation. For instance, while the value that one should be grateful to one's parents, elders and the community at large is present, there is nothing about why one should be grateful? So, the area of logical reasoning is simply not addressed. From the GNH point of view, gratitude, where it is deserved, is a positive sentiment to be used with respect to the monarch and parents. Meanwhile, more complicated concepts like merit and karma are well covered. Rebirth is asserted in history and social studies, but, again, no conceptual grounding of this idea is offered. It is essential to admit that the aspects of merit, karma and rebirth, some of the most complex parts of Buddhist teachings – which one cannot comprehend easily – are often included. It is necessary to supply better arguments, which can be rational and logical, such as those presented by Vasubhandu (Lopen Yignyen). (Not author reference)

In providing biographical sketches of historical figures, both philosophical and religious figures, there is an excessive focus on their magical aspects. The magical aspects have to be contextualised in terms of Vajrayana's technical aspects of Buddhism. Otherwise, it is not comprehensible. It will be like a Theravadan trying to understand Vajrayana; it is not easy to grasp. Some intermediate steps are missing. Pema Lingpa, Phajo, Zhabdrung Rinpoche, and similar kinds of figures are presented in that way. This, I think, carries a great risk. Textbooks have to present them within the Mahayana framework. Tantrism focuses on direct efforts, on the whole person, on energy, on symbolism and on the idea that *samsara* is equivalent to *nirvana.* At least textbooks have to provide the context for understanding Vajrayana figures.

Take this example of a lesson from a social studies textbook. Students are asked to take five scenes on Tertoen Pema Lingpa's ter extraction in Mebartsho and act them out. This is bizarre, because a student cannot enter that magical world and cannot be expected to simulate jumping into a river to find something. The pedagogy is wrong. That is the danger of handling such issues in a trivial way, without any deeper grounding. There are also mistakes with reference to the Pemalingpa, suggesting inadequate grasp of his biography.

Similar criticism can be applied to key references like Sindhuraja and Guru Rinpoche. It might be better to focus on philosophical and ethical ideas in a more simplified form at the lower levels and then advance toward the higher ground. However, it is noticeable that stories and folktales, in which individuals and animals benefit from each other and contribute to human welfare, have been covered. Also, clear examples associated with Ashoka and Buddha are included. Textbooks present these figures as people possessing ethical and spiritual qualities of Bodhisattva. There is a problem of whether they can be understood, especially at the lower level. At the higher level, people have a more complex view of the world. Yet these are not examples of moral figures who can be emulated today. So, presenting them as moral figures will be a bit too elusive and demanding. One would not like to copy Ashoka for one has no institutional power like monarchs, ministers and secretaries. In the case of Pema Lingpa, we live in a world largely different from Pema Lingpa's. So he cannot be copied either.

More pertinent issues need to be considered while portraying moral idealistic figures for students. Just as an example, who the Buddha is and what he taught are two entirely different questions. More could be taught about the ethical and moral relevance of his teachings in schools. In that regard, students should be introduced to the key concepts of the teachings of Mahayana such as *boddichita*, the *boddhisattva path*, principles of interdependencies and non-intrinsic self (emptiness), and skilful means.

Students have two types of memories. One is how things work requiring the development of reasoning and logic. The other is to acquire and remember names and dates. We all learn so many things and, equally, we all forget so many things. We remember only a small proportion of what we have actually learnt. That is declarative memory – to remember dates, people, and names of places. There is too much emphasis on the acquisition of declarative memory by students. Textbooks should be designed to focus less on the mechanical storage of names and dates and should seek to wean students away from rote-learning. But how to reason morally or ethically is currently taught less. It may be fair to argue that one cannot teach this to primary school children; they need to be taught in a different manner. After classes 4 or 5, however, students should graduate towards moral reasoning to deal with existential problems; and on how to approach moral dilemmas. Though there is an abundance of folktales, biographical sketches and mythologies in textbooks, they do not lucidly present moral dilemmas, especially in the context of sustainability.

Good and evil dichotomies are drawn too quickly. Textbooks have to challenge students to think how they would resolve moral dilemmas according to values. In presenting the predicaments of real life, stories could be better selected. Stories jump immediately to classifying things into black and white, right and wrong.

The progressiveness of value education through the classes over time is largely absent. In the lower classes, students should be learning 'do's and don'ts', but that has to change to introduce the concept of right and wrong later. At the final stage, universal values have to be taught. All along, how values are expressed in our culture should be made clear.

Inversion of standards occurs very frequently. For instance, some complex ideas sometimes appear in the textbooks of the lower primary classes whereas they should appear in the upper secondary classes. Meanwhile, some basic concepts occur in the textbooks of higher classes. This reversal of standards occurs quite frequently. This suggests that there is no central consciousness (a mastermind who has a complete and overall view of all textbooks) behind the textbooks. Educationists have to assume that an individual from age six to age fifteen is on a path of improvement, and he or she must be encouraged and motivated to ascend the mountain of knowledge without going over the same ground again and again.

Kohlberg, Piaget and Buddhist moral development theories may help in structuring the progression of standards in values education. It needs to be borne in mind, that, if educationists apply Kohlbergian or Piaget's ideas of moral development for evaluation, individuals brought up in Buddhist moral structures could underscore in Kohlbergian and Piagetian assessment. This happens because self-esteem and competitiveness, for example, are under-emphasised for good reason in Buddhist character development. In values education, it is impossible to escape from Buddhist influences in Bhutan. Most values education is underpinned by Buddhism. If one believes in it, one might want to structure it according to Buddhist ethical development. Buddhist concepts may be unfolded gradually, like a variegated carpet as a student advances through the classes. Moreover, the key concepts of Mahayana like compassion, non-intrinsic existence, interdependencies and ethical training schemes built around ethics, meditation, and wisdom are convergent with universal human development. There is an over-emphasis on adhering to the current law in textbooks. This is conventional and expected, but at a certain point there is a need to distinguish between law and morality. Law is not always moral and vice versa. Law is a medium, not a value in itself.

In a few cases, the standards are haphazard. There is no progression, as mentioned earlier. The gradient in Dzongkha and English for classes 3 and 4 is too steep, seen from the immediate lower level. Its effect may be noticeable in examination results. More students may be failing in class 4 Dzongkha.

There are too many irrelevant asides in civics and history. Histories of the nineteenth and twentieth centuries are taught in classes 9 and 10. They are loaded with details and asides that are not relevant for schools. For example, the treachery of Uma Deywa, a ruler of Bhutan mentioned in one book, is not relevant to the main line of the story. It is relevant only if an individual is doing research to a certain depth and detail. To enable students to see the whole picture, textbooks should not be laden with such details. They should deal with the main focus of the narrative.

In general, textbooks have been reprinted many times since 1992. Three reprints for geography was the minimum and 11 reprints for social science for class 4 was the maximum. The last reprint for social science textbooks was undertaken in 2007. However, no major improvement had taken place since the first print and there had been no review of errors; reprints were produced in a mechanical way. In this respect, the School Curriculum Division of the Ministry of Education could have been a little more discerning.

The final conclusion is that revisions are needed in social science, geography, history, civics and economics textbooks. In the lower classes, science textbooks assume the provision of laboratory equipment. In the absence of equipment, which is a well-known constraint, one is really curious to know how teachers and children cope with this constraint. Much imagination on the part of students and teachers must be involved to compensate for the lack of equipment.

Civics is understandably now outdated. Since 2007 the system of government has changed structurally and institutionally and therefore civics must be revised quickly. Surprisingly, the textbooks in economics are also obsolete. Information stops in the 8th Five Year Plans (FYP's) in economics textbooks. Our students are learning things about 7th and 8th FYPs. Many things have changed in 9th and 10th FYPs. The School Curriculum Division has had no time to revise civics to keep up with events, but in economics, the amount of outdated information is surprising. Most case studies in economics are from *Kuensel* (a national newspaper of Bhutan) issues from 1995, the year the textbook was probably written. Case studies culled from *Kuense*l are 12 years old. For instance, the town of Tsimalakha is depicted in economics as it was in 1995. The transport service characterised is similarly dated. The job prospects and manpower situation described are equally outdated, having been drawn from an old projection by the Royal Civil Service Commission. If we intend to prepare students for the real world of work, serious amendments are needed in economics, civics and history. Too much irrelevant data are being loaded on to our students. For example, the concept of 18 *dzongkhags* in our country is still present in one textbook. A great deal of rewriting work awaits the School Curriculum Division, the Royal Education Council and the Ministry of Education in this respect.

Finally, mistakes found in the textbooks have been quantified, which is a byproduct of checking for values content in the course of reading 148 chapters out of 176 chapters. There is no guarantee that all the mistakes have been detected. The total number of pages read was 3743 and 568 errors that included typos were found in those pages. They do make a huge difference. Sometimes a mistake changes the meaning of a sentence completely. On average one mistake appears every six pages. If we focus on the rate of typo mistakes the density of errors occur more in English, science and social studies textbooks.

There are a large number of texts related to values education: 429 episodes in 148 chapters. In terms of values content, the breadth of issues or episodes covered is adequate. But it is how one can deliver values education that is conceptually clear that needs improvement. If it is dispersed and vague, neither students nor teachers will able to draw conclusions.

5.9 Moral Development

Finally, there is a need to re-assess textbooks in Bhutan with regard to moral development, for which better assessment tools have to be created. Are students developing moral reasoning power? Are students being given entrenched values? Are students developing aesthetic and artistic sensitivity towards moral issues? They should experience a strong reaction, indeed revulsion, against anything that is negative or harmful. Wholesome and positive emotions should be repeatedly encouraged until they become part of their inner fabric and form an essential part of their character. Ekman (2003), an eminent psychologist, demonstrates that when moods arise repeatedly, they become character traits. Character is the consolidation of emotions and moods. Only when moral values and emotions are manifested in behaviour will individuals develop a moral identity. Although students can be inclined towards moral behaviour and their beliefs and attitudes can reflect correct values, teaching in schools will be of little benefit if the broader policies that influence public behaviour are not pro-GNH. The government spends half of the GDP every year and as a result, the government has an enormous influence on the direction taken by the society. Individual values and behaviour are influenced by its policies and programmes. If the agencies of the ministries, para-statal bodies and businesses are sympathetic to GNH, spreading values education will be an easier task.

5.10 Conclusion

"I am a firm believer that if there is one word that will stand out above all other words when we describe our country's amazing journey of modernization over the last few decades – it is Education", said His Majesty, King Jigme Khesar Namgyel Wangchuck, at the 3rd Convocation of the Royal University of Bhutan, 2009. This reaffirms the power of education and learning in realizing the goals of GNH, which is the vision of the country. Schools provide the perfect settings for directly or indirectly inculcating values aligned with GNH values and thus making GNH part of learning experience in the schools. The value basis of Buddhism resonates with values education for GNH. The fundamental Buddhist value orientation propagates values that restrict narrow interests to a reasonable degree on the one hand and commits to values that affirm the welfare of all on the other. If GNH is a development goal of the country, it is imperative ~~is~~ that every citizen, in urban or rural areas, be aware of this and participate in making this vision a reality. Such a wide participation in designing a qualitatively different growth and taking this imagined path is possible through schools.

References

Dewey, J. (1916). *Education and democracy*. New York: Macmillan.

Ekman, P. (2003). *Emotions revealed: Recognizing faces and feelings to improve communication and emotional life.*. New York: Times Books (Henry Holt and Company LLC).

Gallenkamp, M. (2010). *Democracy in Bhutan: An analysis of constitutional change in a Buddhist Monarchy*. New Delhi: IPCS Research Papers.

Government of Bhutan: *The constitution of The Kingdom of Bhutan*. http://www.bhutanaudit.gov.bt/About%20Us/Mandates/Constitution%20of%20Bhutan%202008.pdf. Accessed 8 June 2012.

Kohlberg, L. (1981). *The philosophy of moral development: Moral stages and the idea of justice* (Vol. 1). San Francisco: Harper and Row.

Mathou, T. (2008). *How to reform a traditional Buddhist Monarchy. The political achievements of His Majesty Jigme Singye Wangchuk, the fourth King of Bhutan (1972–2006)*. Thimpu: The Centre for Bhutan Studies.

McDonald, R. (2005). Towards a new conceptualization of gross national happiness and its foundations. *Journal of Bhutan Studies, 12*, 23–46.

Namgyel, S. (2010). Reorienting classroom interaction. Seminar: *The pursuit of happiness – Bhutan's strategy of promoting well-being*. New Delhi: Malvika Singh.

Piaget, J. (1932). *The moral judgement of the child*. Glencoe: The Free Press.

Sandel, M. (1998). *Liberalism and the limits of justice*. New York: Cambridge University Press.

Shiva Kumar, A. K. (2010). Happiness matters. *Seminar: The pursuit of happiness – Bhutan's strategy of promoting well-being*. New Delhi: Malvika Singh.

UNESCO. (2007). *Education for sustainable development: Linking learning and happiness*. Bangkok: UNESCO.

Ura, K. (2009). *A proposal for GNH value education in schools*. Thimphu: Kuensel Corporation Limited.

Ura, K. (2010). Measuring the happiness. Seminar: *The pursuit of happiness – Bhutan's strategy of promoting well-being*. New Delhi: Malvika Singh.

Varma, P. (2010). The problem. Seminar: *The pursuit of happiness – Bhutan's strategy of promoting well-being*. New Delhi: Malvika Singh.

Chapter 6
Environmental Education in India – The Shifting Paradigm

Shailaja Ravindranath

6.1 Introduction

Environmental Education (EE) has grown so vast. It has embraced so many concepts, ideas and practices, since its emergence in the 1970s, that it is increasingly becoming difficult to define its boundaries.

> "The forest nurtures us like a mother; you will only be able to use your axes on it, but you have to use it first on us" (Guha 1989: 15) declared Gauradevi, one of the prominent women leaders of the famous Chipko movement in the Uttaranchal region in the Himalayas in 1974. Using the protest method of nonviolence, local people, mostly women, calling out slogan: "What do the forests bear? Soil, water and pure air. " (Shiva 1988: 77) formed a circle around the trees in the Himalayan forests protecting them from men who wanted to axe them down. Spreading the awareness, "this forest is the source of our livelihood. If you destroy it, the mountain will come tumbling down onto our village" (Bankoti 2008: 113), Gaura Devi mobilized the women of the region. Sparked off as a protest against the tree felling in Himalayas, Chipko ('to adhere' in Hindi, the official language of India), the popular ecological movement in Independent India, became instrumental in bringing a 15-year ban on tree felling in the region in 1980 by Indira Gandhi, the then Prime Minister of India (Shiva and Bandyopadhya 1986: 135). It generated pressure for a people-centred ecologically sensitive natural resource policy.

This constructive movement also gave birth to new beginnings in the areas of environmental awareness and education throughout India spreading the movement to several states across the country. For instance, following Chipko, the 'Hug the Trees' movement (*Appiko* in the local language Kannada) in Karnataka, one of the Southern States, used fascinating, highly creative and unique techniques such as foot marches, slide shows, folk dances, street plays, to raise awareness to halt the

---

S. Ravindranath (✉)
Regional Director, Southern Regional Cell, Centre for Environment Education (CEE), Bangalore, India
e-mail: shailaja.ravindranath@ceeindia.org

R.N. Gorana, P.R. Kanaujia (eds.), *Reorienting Educational Efforts for Sustainable Development*, Schooling for Sustainable Development,
DOI 10.1007/978-94-017-7622-6_6

tree felling in Western Ghats region and redefine development for a sustainable future (Hegde 2010).

Chipko, Appiko and several such environmental movements in India, kindled by the people most affected, summarize the essence of EE. "Environmental Education should simultaneously attempt to create awareness, transmit information, teach knowledge, develop habits and skills, promote values, and provide criteria and standards, and present guidelines to problem solving and decision making. It therefore aims at both cognitive and affective behaviour modification". "This is an action oriented, project centered and participatory process leading to self-confidence, positive attitudes and personal commitment to environmental protection. Furthermore the process should be implemented through and interdisciplinary approach" (UNESCO-UNEP 1992: 4).

Perceived as 'Education about the environment, through the environment and for the environment', EE aims at improving environmental quality and thereby the quality of life of humans and other life forms as well. It is comprehensive, continuing, and interdisciplinary in nature and is related to almost all subject areas. In response to the growing concern for the deteriorating state of the environment, EE emerged during the United Nations Conference on 'Human Environment' (1972) in Stockholm, and has widened in its scope since then. It has embraced the concept of sustainability with multiple perspectives in the 1990s after the United Nations Conference on Environment and Development (UNCED) held at Rio in Brazil in 1992. EE is undergoing a major transformation following the World Summit for Sustainable Development (WSSD 2002) in Johannesburg, re-emphasizing the need for adapting sustainable practices. This has resulted in the growth of the concept of education for sustainable development (ESD) in recent years.

ESD addresses three key areas of sustainable development: society, environment and economy. The inspiring words of Chico Mendes (Brazilian Rubber tappers' leader and environmentalist), "At first I thought I was fighting to save rubber trees, then I thought I was fighting to save the Amazon rainforest. Now I realize I am fighting for humanity" (http://www.uwcsea.edu.sg/page.cfm?p=1088. Accessed on April 2, 2013), explains the difference between EE and ESD precisely. The United Nations declared the decade 2005–2014 as the Decade of Education for Sustainable Development (DESD) in recognition of the central role of education and communication in sustainable development. From EE to ESD, it has been four decades of long journey. Where does India feature in this journey?

6.2 Environmental Education in India

The diverse landscapes in India ranging from deserts to coasts, extensive mountain ranges to vast plains, support a wide range of socio-cultural aspects associated with biological diversity and ecological services. Traditional societies living in these diverse landscapes of the country perceive conservation as a part of their religion, culture and education for sustainable lifestyles. For instance, forests, tree groves, certain plant species like Peepal (*Ficus religiosa*) and animal species like Black Buck (*Antilope cervicapra*) receive special reverence because they are important in maintaining the tropical biodiversity. This respect is further extended to nonliving entities of the environment such as water bodies and mountains. For instance, Ganga, the river, and the Himalayas, the mountain ranges, are worshipped for their crucial ecological services. Attaching divinity to these various components of the environment ensured their protection, especially in the pre-industrial era.

Evidence of such traditions and practices can be seen even today in several pockets of the country. The green patches in and around the habitations of the Bishnoi community in the western part of India, for instance, stand testimony to their continued practice of vegetarianism, non-violence, protection of trees and respect for all living things stemming from deep rooted knowledge about habitats and ecosystems. The magnificent traditional water harvesting systems, fascinating sacred groves, some of which still function, illustrate the knowledge of traditional societies about the effective and sustainable management of natural resources. The Indian subcontinent is considered as one of the greatest repositories of ethno-biological knowledge.

Traditional societies have also expressed concerns about the impact of environmental degradation and the need for conservation for human survival. These concerns and practices reveal the highest degree of awareness and knowledge these societies possessed about the environment that they were living in. Environmental concerns and practices resulting from EE, hence, are not a new idea or concept in India. This is exactly what Mrs. Indira Gandhi, the then Prime Minister, expressed in 1988 while launching the World Conservation Strategy in India, "The interest in conservation is the rediscovery of a truth well known to our sages. The Indian tradition teaches us that all forms of life: animals and plants are so closely linked that disturbance in one gives rise to imbalance in the other" (http://www.ceeindia.org/newintra/papers/Strategies%20in%20Environmental%20Education.htm. Accessed on April 2, 2013).

However, the increase in population, rapid modernization (for instance, replacement of local efforts and community water management practices with large dams for irrigation and hydroelectricity generation), changing lifestyles and livelihoods, altered socio-cultural systems, globalization and new markets over the past century, have dissociated humans from nature. Traditional wisdom, knowledge systems and practices are fast eroding. As a result, education, awareness and value systems towards environment, which were passed on to generations through practice and oral traditions, have lost their significance. This knowledge base is neither considered important, nor has it been carefully integrated into planning and policy of the present national governance which largely follows the West. This is evident from the mounting pressure on natural resources like water, air and forests, and unprecedented environmental degradation across the country. EE therefore needs much more attention and emphasis in the present scenario to re-establish the harmonious relationships between people and nature. This will also promote changes in attitudes and practices of the people and the government, and redefine development for a sustainable future.

6.3 How Is India Preparing Itself to Take on These Challenges?

In the later part of the 1850s, during British India, concerns of Lord Dalhousie and Lord Canning about the effects of deforestation and environmental degradation led to the formulation of the Indian Forest Policy. The Forest Charter of 1855 and the Indian Forest Act of 1878 gave a sound scientific base for forest conservation. Modern EE in India perhaps has taken its roots from this Act, as argued by Sinha (http://www.sfri.org/images/general/article_brit_ind.pdf. Accessed on April 2, 2013). Post-Independence India has made significant strides in rediscovering and promoting EE across the country.

The Indian Constitution (coming into force in 1950) is amongst the few in the world that explicitly makes environmental conservation "a fundamental duty of every citizen" (http://india.gov.in/my-government/constitution-india/constitution-india-full-text. Accessed on May 12, 2012), and guides the national government and the people through their actions. For instance, Article 48A in the Directive Principles of the State Policy (http://india.gov.in/my-government/constitution-india/constitution-india-full-text. Accessed on May 12, 2012) declares that "the state shall endeavour to protect and improve the natural environment so as to safeguard the forests and wildlife of the country". Article 51A(g) in "Fundamental Duties" (http://india.gov.in/my-government/constitution-india/constitution-india-full-text. Accessed on May 12, 2012) imposes a similar responsibility on every citizen, "to protect and improve the natural environment including forests, lakes, rivers and wildlife and have compassion for living creatures". Article 253 of the Constitution empowers parliament to legislate on all matters linked to the preservation of natural resources.

It is interesting to note that for the first time problems related to environment received attention in India's fourth Five Year Plan (1969–74) (http://planningcommission.nic.in/plans/planrel/fiveyr/welcome.html. Accessed on April 2, 2013). Since then, India's Five Year Plans have emphasized environmental awareness and education with financial outlays. Five Year Plans are India's economic plans. The Planning Commission develops, executes and monitors these economic plans once in 5 years and assesses the country's resources and formulates development plans that lead to the most effective utilization of resources. The first Five-year Plan was launched in 1951 and the eleventh plan (2007–2012) was completed in 2012.

The turning point in environmental awareness and management in India came in 1972 during the United Nations Conference on Human Environment at Stockholm. Mrs. Indira Gandhi was the first Head of State to point out the connection between environment and poverty in this Conference:

> …environment cannot be improved in conditions of poverty, unless we are in a position to provide employment and purchasing power for the daily necessities of the tribal people and those who live in and around our jungles, we cannot prevent them from combing the forests for food and livelihood; from poaching and despoiling the vegetation. How can we speak to those who live in villages and slums about keeping the oceans, the rivers and the air clean when their own lives are contaminated at the source? (http://www.iisc.ernet.in/insa/ch33.pdf. Accessed on April 2, 2013).

The Government of India has a two-tier structure: a federal structure with elected Governments in states. The federal structure, also called the Union Government or Central Government, is the governing authority of the Federal Union of twenty-eight States, six Union Territories (administered directly by the Central Government) and a national capital territory (NCT-Delhi). State Governments follow the guidelines and policies of the Centre in many aspects, including education, making necessary adaptations based on the people, culture, resources and so on of the respective state. Following this Conference, the Government of India recognized the need for integrated environmental planning for the first time and formally constituted the National Committee on Environmental Planning and Coordination (NCEPC) in 1972. NCEPC became the National Committee on Environmental Planning (NCEP) in 1981 and gradually evolved as a separate Department of Environment and reached the fully fledged status of Ministry of Environment and Forests (MoEF) by 1985.

EE grew stronger in India following the first Intergovernmental Conference on Environmental Education in Tbilisi (1977), which is a landmark in the history of EE. Realizing that education is the panacea to halt the rapid environmental degradation in India, the 6th (1980–1985) and 7th (1985–1990) Five Year Plans laid an emphasis on EE for creating and intensifying environmental awareness at all levels across society (http://planningcommission.nic.in/plans/planrel/fiveyr/welcome.html. Accessed on April 2, 2013). Using both non-formal and formal channels and through the relevant curricula, publications and training, the plans were intended to stimulate public participation in environmental protection. (See Box 6.1)

The Ministry of Environment and Forests (MoEF), soon after its formation, launched the nation-wide Environmental Awareness Campaign (NEAC) in 1986. This was designed to create awareness among the people and inspire them to participate in the protection of the environment. There was an overwhelming response from teachers and NGOs with more than 10,000 participating in the first campaign across the country. The programme addressed local environmental problems and hence became very popular and successful in reaching out to people. This campaign made EE more visible in the country which otherwise was limited to sporadic experiments (See Box 6.2).

In the late 1980s, there were several interesting experiments undertaken to bring EE into the formal education system. These included teacher training programmes, campaigns, environment day celebrations, publishing educational materials on environmental concepts and issues, and carrying out environmental monitoring in institutions like schools and colleges. The financial support for these experiments came from the MoEF, Ministry of Human Resource Development (MHRD), State Governments and international donor agencies like NORAD. The National Policy of Education (NPE 1986) states, “There is a paramount need to create consciousness of the environment. It must permeate all ages and all sections of society, beginning with the child. Environmental consciousness should inform teaching in schools and in colleges. This aspect will be integrated in the entire educational process” (http://mhrd.gov.in/sites/upload_files/mhrd/files/NPE86-mod92.pdf. Accessed on March 5, 2013). MHRD which holds the education portfolio at the Central level, acknowledging the importance of EE in shaping knowledge, skills and values

Box 6.1: Environmental Education, Training and Awareness
18.54. For creating and intensifying environmental awareness at all levels of Indian society, both formal and non-formal educational channels will have to be utilized. The basic thrust of the programme would encompass the following objectives:

- Incorporation of environmental themes in educational curricula and teaching aids/materials in the formal environmental education sector.
- Encouraging non-governmental organizations, mass media and concerned organizations to promote non-formal education.
- Providing aid to professional societies and institutions for environmental education activities.
- Developing the National Museum of Natural History and establishing Regional (Satellite) Museums.
- Feasibility studies for establishing a National Institute for Environmental Management.
- Promoting the setting up of Interpretive Centres in national parks, botanical gardens, zoos and biosphere reserves.
- Promoting manpower development and generating literature and audio-visual material for environmental education.
- Establishing Centres of Excellence in Environmental Education and Research.

18.55. In the formal education system there will have to be the fullest involvement of the Ministry of Education, and in particular the NCERT (for schools) and UGC (for universities). It is proposed to arrange for comprehensive training and consultancy services, besides having facilities for research in environmental management, on aspects of relevance to corporate executives, senior planners and administrators. (7th Five Year Plan, 1985–1990, Vol 2; 18.54 and 18.55) (http://planningcommission.nic.in/plans/planrel/fiveyr/welcome.html. Accessed on April 2, 2013)

among students, has been instrumental in bringing out the this policy as early as 1986. The trials on EE however increased during the following National Policy of Education. MHRD through its autonomous body – the National Council for Educational Research and Training (NCERT) which is the apex body in the area of developing curriculum for the formal school education – attempted to infuse environmental concepts and values of environmental protection into textbooks. Instructional materials, classroom transactions and co-curricular and extra-curricular activities both in science, social sciences and even in languages carried environmental issues. This was the beginning of mainstreaming of EE in formal education. Some of the states like Karnataka, Maharashtra followed this pattern. However, these efforts were highly inadequate and did not make much impact on mainstreaming of EE in formal education.

Box 6.2: Moving Towards Visibility

National Environmental Awareness Campaign, popularly known as NEAC launched as early as 1986–1987, is a flagship programme of the MoEF. The objective of the programme is to enhance awareness among a wide spectrum of the population on various environmental issues and encourage them to participate in the protection of the environment. Every year the Campaign focuses on a theme. Through NGOs, scientific and educational institutions; using an array of activities like rallies, door to door campaigns, demonstrations, and the campaign focuses on environmental issues and problems relating to the main/regional/local themes and reaches out to people. Today, after almost 25 years since the NEAC launch, the programme continues due to the overwhelming positive response it receives. Through an established network of 12,000 agencies, mainly NGOs, the programme has reached out to various sections of society across the country through more than 2, 00,000 projects in 25 years. 34 Regional Resource Agencies (RRAs) service and facilitate this network. MoEF funds nearly 8000 programmes every year across 28 States and 7 Union Territories in the country. (CEE 2011: 5–6)

In 1991, following the petition of M.C. Mehta, a lawyer of the Supreme Court and an environmentalist, the Supreme Court stated that EE should be made compulsory in formal education system. The Supreme Court is the apex court of the country that has original appellate and advisory jurisdiction. The Constitution of India gives an extensive original jurisdiction to the Supreme Court with regard to enforcement of Fundamental Rights. This posed several challenges such as how to mainstream EE. Should EE be a part of the taught subjects or should it be a separate subject? How can teachers be trained to teach EE since most of the teachers have no experience in EE transaction?

The credit for giving direction to the implementation of EE in schools should, to some extent, go to the District Primary Education Programme (DPEP). This is a major educational initiative of the country to supplement the efforts of the Central and State Governments for Universalization of Primary Education (Education for All) in India (DPEP 1994) (www.educationforallinindia.com/page81.html. Accessed on April 25, 2012). DPEP started in 1993–1994 with funding from states, national and international sources – with the major portion of the funding coming from the World Bank.

Through this programme, some of the states attempted to use the environment as a binding and over-arching subject at the primary level (classes 1–5). These trials were based on the encouraging results of some of innovative experiments like Nali-Kali ((Joyful Learning) in EE planningcommission.nic.in/reports/sereport/ser/seeds/seed_edu.pdf. Accessed on April 2, 2013). Integrating the concepts of science and social science concepts, these states brought out what is known as 'Environmental Studies', EVS in short, for the primary classes. An activity-based approach was introduced and teachers were trained in the methodology. This experiment only got bigger and better. Impressed by the results, NCERT, in its National Curriculum Framework for School Education (NCFSE) 2000, recommended EVS as a subject for the primary level, thus giving 'Environment' it's due recognition as a subject. EVS then was introduced in classes I–V. From classes VI to X environmental concepts and information were infused suitably in science, social science and language textbooks.

In 1999–2000, a massive study was undertaken by the Pune-based Bharati Vidyapeeth Institute for Environmental Education and Research (BVIEER) on the 'Status of infusion of environmental concepts in school curricula and the effectiveness of its delivery' sponsored by the MoEF and the World Bank. The study undertaken in all the states and Union Territories revealed major gaps in the infusion of environmental concepts in the textbooks. Some of the major problems pointed out were a lack of coordination, continuity, concept levels, and inappropriate pictures (BVIEER 2002). This analysis paved the way for the 'greening' of the textbooks at the national level in a phased manner. Many State Education Departments joined hands with MoEF in a project called Environmental Education in the School System (EESS) (2002–2008). With the support of CEE, various environmental concepts were infused in science, social science and language textbooks of classes VI, VII, VIII during 2003–2004. These textbooks were field tested in 100 schools in each state (MoEF). However, this initiative had to face several constraints like many State Education Departments were not convinced; MHRD, the body responsible for mainstreaming EE, did not show much enthusiasm and hence there was no active participation; the process itself was very slow. The project was further taken to ten more states as Strengthening Environmental Education in School Systems (StrEESS).

Meanwhile, the Supreme Court, displeased with the noncompliance of the earlier order of 1991 by the states to make EE compulsory, sent a Directive to the States in 2003, "…We would require every State Government and every Education Board connected with education up to matriculation stage or even intermediate colleges to immediately take steps to enforce compulsory education on environment in a graded way. This should be so done that in the next academic year there would be compliance with this requirement" (Case 865) (http://old.downtoearth.org.in/html/sc_Directive.htm. Accessed on April 4, 2013), to make EE mandatory at all levels of formal education. Failing this, the Supreme Court warned the states that they have to pay a penalty. In 2004, NCERT even brought out a syllabus on environmental studies following the instructions of the Supreme Court. This Supreme Court Directive (2003) accelerated the process of mainstreaming EE.

Concurrently, NCERT, as a part of its mandate published the National Curriculum Framework (NCF) in 2005 and environmental education appeared as 'Habitat and Learning' (NCERT 2005a, b: 64). (The National Curriculum was reviewed by twenty-one Focus Groups. One of them was on Habitat and Learning. The Focus Groups brought out Position Papers which became part of the National Curriculum

Framework package.) NCF also recommended the infusion of environmental issues in different disciplines for the greater impact, ensuring innovative EE methods and adequate time. The Centre and the states have prepared new syllabi based on NCF 2005. They have also initiated the process of 'greening' the textbooks based on the new syllabi in a phased manner and the process is expected to be completed by 2015, in all the twenty-eight States and seven Union Territories, and will be introduced in the respective classes. As in NCFSE 2000, NCF 2005 also has recommended environmental studies (EVS) as a subject for classes I–V. In higher classes (classes VI–X) environmental concepts and issues have to be infused in science, social studies, languages and mathematics. In classes XI and XII environmental concepts will have to be integrated into various subject areas. It is introduced as an optional subject and through projects as well.

Environmental Education had not received adequate attention in higher education until the Supreme Court Directive in 2001. The Supreme Court directed the University Grants Commission (UGC) to introduce a basic course on environment at every level in higher education. Complying with the direction, the UGC has formulated a "compulsory" course of a 6 months core module in environmental studies across the country for undergraduate courses in all branches of higher education, thus giving environment its space and time in academic programmes (http://www.ugc.ac.in/oldpdf/modelcurriculum/env.pdf. Accessed on May 12, 2012). UGC has also brought out a textbook 'Environmental Studies for Undergraduate Courses' covering a range of topics. It aims at creating "pro-environmental attitudes" and changing behavioural patterns and lifestyles of youth (http://www.ugc.ac.in/oldpdf/modelcurriculum/env.pdf. Accessed on May 12, 2012). It has notified all 563 (figure as of 2012) (see http://en.wikipedia.org/wiki/List_of_universities_in_India) universities to use this course material.

Many universities have introduced the UGC module in their respective colleges as a common course for undergraduate students of all disciplines such as arts, science, commerce and law since 2005. Technical universities and colleges, such as engineering, management, and agricultural sciences, have introduced environmental studies as a compulsory subject of study in their respective programmes.

India is perhaps one of the earliest nations to have internalized EE (Environmental Education) in its programmes and processes. The growth graph of EE in India thus can be traced alongside international thinking. EE has come a long way. The scope of EE has broadened in India along with the educational and societal developments, and needs over these past four decades from its beginnings in the 1970s. While it has taken almost 30 years (from the UN Conference on Human Environment in Stockholm in 1972 to the Supreme Court Directive in 2003) for EE to carve a niche within the mainstream education (formal education) and educational policy, the concept of ESD seems to be getting integrated at a faster pace. This may be because the Central and State planning processes are laying greater emphasis on the concept of sustainable development and equitable growth. Another reason probably could be that in India most of the EE initiatives have always focused on the social and economic issues as well. From the outset the transition of EE to ESD appears to be a smoother affair. However, the depth of understanding the scope, the processes and the implications of ESD is still debatable.

6.4 Major Environmental Education Initiatives and Key Players

"Environmental issues will remain a serious concern in India…For the future growth to be rapid and sustainable, it needs to be as resource-efficient and environmentally benign as possible" states India's Vision 2020 (http://planningcommission.nic.in/reports/genrep/pl_vsn2020.pdf. Accessed on April 2, 2013). Keen on becoming a resource efficient country, it is clear that India is promoting several green actions. Several government organizations, non-governmental organizations, academic and research centres, universities, communities and individuals, corporate sector, and mass media are putting their efforts into this endeavour with major initiatives and funding coming from the Government of India. Several ministries and government departments at the Centre and State levels such as agriculture, new and renewable energy sources, urban development, rural development and Panchayati Raj (the grass-roots units of self-government), water resources, tourism have included environmental awareness and education as an important step towards this vision.

6.5 Government Initiatives

The Ministry of Environment and Forests (MoEF), now Ministry of Environment, Forest and Climate Change (MoEFCC) since 2014 considering the comtemporary issues related to climate change, is the leading Ministry and a major funder of EE. It has been emphasizing the "creation of environmental awareness among all sectors of the country's population" as one of the key tools to plan, promote, coordinate and intensify the implementation of various environmental and forestry programmes since its formation (http://envfor.nic.in/mef/intro.html. Accessed on March 10, 2013). Realizing the importance of an environmentally literate society, the MoEFCC has been conducting large and ambitious EE programmes at different levels including students and the general public, in order to make EE a people's movement. These initiatives use various methods and approaches like organizing seminars, campaigns, setting up of eco-clubs, and the production of awareness materials. For instance, it coordinates the National Environment Awareness Campaign (NEAC) and the National Green Corps (NGC), the two flagship programmes on EE in India (National Environment Awareness Campaign)

(See Boxes 6.2 and 6.3)

Box 6.3: National Green Corps (NGC) – A Decade of Green Action
Realizing the tough job of changing the attitudes of over a 100 core people, MoEF has prepared a National Green Corps of students to do the job. National Green Corps (NGC), initiated by the Ministry of Environment Forest and Climate Change (MoEFCC), Government of India in 2001 is the largest network of over 1.2 lakh schools. Working through the school eco clubs, NGC promotes green actions through the students for positive impacts on the environment, society and economy. This landmark initiative has inspired more than 60 lakh young students to spearhead the green movement across the country for ensuring a sustainable planet. NGC has built a cadre of young people equipped with scientific knowledge, skills and positive attitude to conserve the environment; promoted school-society interaction, sensitizing the society thereby. With the growing number of eco clubs in all the States and Union Territories (UTs), NGC has achieved significant success in 10 years since its inception in 2001 (CEE 2010: 5)

Besides these programmes, MoEFCC has established seventy-six Environmental Information System (ENVIS) centres to collect, collate, store, retrieve and disseminate information on various environmental themes and issues including environmental education at the national level. (http://www.envis.nic.in/centres.html. Accessed on March 10, 2013). It has established the Centre for Environment Education (CEE) in 1984 at Ahmedabad and the C.P. Ramaswamy Environmental Education Centre (CPREEC) in 1988 at Chennai as 'Centres of Excellence' in EE to play a pivotal role in setting the pace and the agenda for EE in the country. Recognizing the vital role of environmental conservation in sustainable development, states have set up a separate department of Ecology and Environment with encouragement from the MoEFCC, to spread and intensity EE in their respective states, in their conservation-related functions, facilitate and coordinate awareness and education programmes on environment and ecology through their various boards and district level centres. As recommended in the 7th Five Year Plan, MoEFCC has also established the National Museum of Natural History (NMNH) in Delhi, a unique facility to promote non-formal environmental education. Regional Museums of Natural History (RMNH) have also been established in Mysore, Bhopal and Bhubaneshwar. Through facilities like zoos and botanical gardens, interpretation centres in national parks and wildlife sanctuaries, MoEFCC is facilitating the processes of creating environmental awareness among the general public with the support of states. It is attempting to synergize the activities of its various centres. For instance, the Centre for Environment Education (CEE) collaborated with Central Zoo Authority (CZA) and developed the Master Education Plan for Zoos of India in 2009 in order to provide a framework to plan active educational programmes as part of Zoo activities.

The Ministry of Human Resource Development (MHRD) is another key player and major funder augmenting the country's efforts in EE. This ministry, in-charge of planning, implementation and monitoring of education in the country with support from the State Education Departments, has been working to mainstream EE in formal education. Besides this, MHRD, through Environmental Orientation to School Education (EOSE), a scheme to enrich the pool of locale-specific EE materials in local languages, which is the key to spread the message, had brought out state relevant materials on EE with the help of NGOs, for students and the general public.

While MoEFCC targets the general public (non-formal system), and uses a co-curricular and extra-curricular approach to EE while dealing with students in the formal system, MHRD plays a crucial role in bringing a curricular framework to EE in the formal education system that leads to mainstreaming by providing a large structure for integrating EE and ESD. This important synergy between the two ministries is ideal for spreading ESD across the country, addressing almost all the target groups.

6.6 Research and Academic Institutions

There are several academic and research institutions working to strengthen environmental awareness in the country, such as: the Bombay Natural History Society (BNHS), Mumbai; the Centre for Ecological Sciences (CES) at the Indian Institute of Science (IISc), Bangalore; the National Environmental Engineering Research Institute (NEERI), Nagpur; the National Institute of Oceanography (NIO), Goa; National Institute of Ocean Technology (NIOT), Chennai; Forest Research Institute (FRI) Dehradun; and various university departments.

6.7 Non-governmental Organizations

One of the strengths of India's environmental education movements is the vast network of Non Governmental Organisations (NGOs). Over 10,000 NGOs and community based institutions at national, regional and state levels are actively engaged in improving awareness on environment and development issues (See Box 6.4). Recognizing the contribution of NGOs at various levels in spreading environmental awareness, the Government is building a number of partnerships to achieve the tasks ahead. NGC is a good example of such joint ventures with the participation starting at the grass root level – from schools, community, district administration – to State Departments and NGOs with the Central support, the strengths of each complementing the other.

Box 6.4: Awakening Public Consciousness to Justice
The Coastal Zone Regulation is a notification served by MoEF in 1991 in order to conserve the coastal ecosystem and to protect the livelihood of the traditional fishers in India. However, realizing the inadequacy of the notification in addressing several coastal issues, including that of disasters following tsunami in 2005, MoEF proposed a Coastal Management Zone (CMZ, 2008) in line with the internationally recommended integrated coastal zone management. Fearing that CMZ 2008 may hamper the livelihood of traditional fishers, the NGO machinery was quick to spread awareness among the coastal communities about the impacts of CMZ 2008. Sensing public unrest, MoEF commissioned CEE to hold public consultations to collect the view points of the coastal communities on CMZ 2008. CEE made the notification available in the local languages before the consultations to facilitate public participation. Powered by the knowledge, people participated in large numbers to communicate their not-so-favourable views on CMZ, 2008. Based on the public opinions supported by the report of the parliamentary committee, MoEF allowed CMZ 2008 to lapse on July 22, 2009 with the recommendation to strengthen the CRZ 91 (http://envfor.nic.in/divisions/iass/2009-07-24%20Press%20Release%20-%20CMZ%20Lapse.pdf. Accessed on March 6, 2012).

6.8 Communities Energized by Environmental Awareness

Traditional communities, in particular, have always contributed to environmental conservation because they are aware of the fact that the environment is their lifeline. Even today, people in several regions are working towards conservation of water, agro-biodiversity, sacred groves, mangroves and so on because of the immense knowledge they possess. An example is the Kolar district and the magic that environmental awareness spelt on the women there (See Box 6.5).

Communities are also becoming aware of the adverse impacts of developmental projects. This is resulting in peoples' campaigns to pressurize the government and influence policies for environmentally sound and sustainable development practices. The world famous 'Save Silent Valley Movement' is an example of how communities can turn the decisions with the strong tools of information and environmental education (See Box 6.6).

These 'green' movements, powered by the awareness that natural resources are fast eroding, have embraced a wide spectrum of issues from forest conservation to pollution abatement. Initiated at the grass roots, including local communities, indigenous communities and women, NGOs have given voice to their cause and supported them to reach out to decision makers. Interestingly, most of these movements

Box 6.5: Community at work

Kolar district is a dry zone of Karnataka, in Southern India. Short-sighted policies have encouraged over exploitation of groundwater in this historically draught prone area. The resulting water poverty forced Gram Vikas, a local NGO take up tank restoration with its strong network of women Self Help Groups (SHG). Besides physical cleaning of tanks, SHGs undertook a 'Jatha' on tank restoration and ecological advocacy. Women visited 400 villages in 3 Taluks and through songs and slogans, organizing theme conferences, inviting policy-makers and media drew their attention to the issue. The strategy worked very well and a 'Tank Development Committee' was initiated in each village they visited. Realizing the ordinary women's extraordinary commitment, the State Government decided to desilt tanks of the State. With this was born the Tank Restoration Project 'Jala Samvardhane Yojana (JLSY)' in 2002 funded by the World Bank. Awareness did empower women to protect their water resources. (Acharya 2006: p. 49–58)

Box 6.6: The Man and the Monkey Debate

The Kerala State Electricity Board (KSEB) proposed a Hydro-Electric Project (SVHEP) in Silent Valley, thick, lush, evergreen forests in the State of Kerala in the South Western part of India. As a part of this project a dam was planned across Kuntipuzha, a major river that originates in the Silent valley. Large number of trees was cut for the purpose in the early part of 1970s. Silent valley being the only home of lion–tailed macaque (*Macaca silenus*), scientists and researchers raised their concerns over the project. The concerns later were translated into heated environmental debate of the decade. NGOs, scientists, activists, poets joined this fiercely Save Silent Valley movement with Kerala Sasthra Sahithya Parishad (KSSP) taking the lead. They raised public awareness about silent valley and its importance through reports and lectures, debates effectively. Infuriated people of Kerala filed a petition before the High Court of Kerala, against the forest cutting in the silent valley area. The court banned the clear cutting though, it lifted the ban in early 1980s. But, Mrs. Indira Gandhi, the then Prime Minister of India, requested the Government of Kerala to stop the tree felling. The project was abandoned in 1983 based on the report given by the multidisciplinary committee headed by Prof. M. G. K. Menon that the project could cause a serious, irreparable ecological damage. The Government of Kerala declared the Silent Valley area, a National park in 1984. The debate on 'the man and the monkey' still continues… (CEE 2002: p 61–64).

have yielded the desired results. Realizing the strength of people, the governments at the central and state level have recognized communities as an integral part of developmental projects making their participation crucial. Hence environmental awareness and education have become important components in all the programme areas.

6.9 Corporate Contributions

In the corporate sector, there are few players investing in EE. Small initiatives in schools have come from foundations like Arcelor Mittal, Sir Ratan Tata Trust, BOSCH and Infosys. The recent Azim Premji University in Bangalore is another corporate initiative which focuses on teaching and research in ecology and environment. Some banks and industries, under their Corporate Social Responsibility (CSR), give small funds for one-off events like competitions, campaigns, rallies, celebration of environmentally important days. The central Government is also finding newer ways of working with business establishments who wish to contribute to the body of EE. For instance, the Paryavaran Mitra (Friends of Environment) programme is a 3-sector partnership with government (MoEF), NGO (CEE) and corporate body (ArcelorMittal).

6.10 Media Interventions

Mass media, such as print, TV and radio, in various parts of the country occasionally cover environmental issues. For instance, Greenathon of the NDTV-Toyota is a nationwide green campaign which includes raising funds to save the environment. This campaign with its 24 h live telecast across the NDTV network includes music, live chats, green activities, green pledges on environment, reaching out to a large section of society in a short time. On environmentally important days like World Environment Day and Earth Day, the print media, in particular, bring out special supplements on various environmental issues. There are some documentaries and films made on environmental issues. Recently, *Wild Dog Dairies* (2010), a film made by wildlife photographers Krupakar-Senani, shot in the parts of protected areas of southern India, has won Green Oscars.

6.11 International Commitments

India's position is vital in the international scenario in terms of contributions to international policies and treaties. It is also is a significant recipient of technologies and finance. As an active participant in several international meetings, conferences and negotiations, India is party to several important international treaties and commitments related to environment and development such as Agenda 21, the Montreal Protocol, the Millennium Development Goals (MDGs), the United Nations Framework Convention on Climate Change (UNFCCC), and the Convention on Biodiversity (CBD) and now the recent Paris Climate Agreement (April, 2016), showing its commitment towards environmental protection and sustainable development. However, all these commitments require a strong educational intervention, particularly ESD which India is working towards. For instance, in this Decade of ESD, India has taken major strides to strengthen existing educational programmes like mainstreaming EE in formal education, biodiversity education. MoEFCC along with the Ministry of Railways has initiated a unique project to spread awareness about the biodiversity across the regions of the country through a travelling exhibition on a train with the help of CEE and other research and academic institutions. India also has initiated new educational programmes on coastal and marine conservation, climate change to strengthen the National Action Plan on Climate Change (NAPCC). CEE has brought national and international experts of ESD together on a common platform through conferences to share experiences and resources for the promotion of ESD across the country, such as the 'Education for a Sustainable Future' (ESF) – the first international event to mark the UN Decade of Education for Sustainable Development (DESD) in 2005; 4th International Conference on Environment Education in 2007; Ethical Framework for a Sustainable World, to mark 10 years of Earth Charter in 2010. Several international agencies, particularly UNESCO, UNDP, UNEP, provide resources for the promotion of ESD in the country.

6.12 The Challenges for India

It is obvious that India is taking a great deal of interest in making the country clean and green. However, are these efforts bringing the desired results? Perhaps not. The impact of EE is slow and hardly visible. Where is the problem? The strengths and the weaknesses of EE initiatives indicate the challenges that the country will face in the coming years.

6.12.1 Strengths

- India's rich traditional practices are environmentally sound and sustainable.
- The Indian constitution makes environmental conservation a duty of every citizen and state.
- The National Policy of Education (1986) stresses the need for EE for all sections of society.
- The Supreme Court's Directive (2003) has made EE compulsory at every level of formal education resulting in the mainstreaming of EE.

- The Central Government provides financial and technical assistance for EE to states.
- All State Government departments recognize EE as an essential part for environmental conservation and sustainable development. Many states have even provided financial outlays to EE.
- Many key players including academic and research institutions are actively participating in awareness promotion, skills up-grading in environmental conservation and they also fund EE activities.
- The Supreme Court has passed a Directive that the media must show a certain amount of programmes to create environmental awareness, free of cost.
- India is shifting its paradigm from EE to ESD as it considers ESD as a major driver of change in achieving sustainable development.

6.12.2 Weaknesses

- Defining the scope of EE and ESD has been a major limitation. Lack of understanding of the depth of EE/ESD has restricted its scope and significance. This has made EE/ESD more ornamental in application; limiting it to lectures, tree planting, garbage management, competitions.
- In the enthusiasm of valuing EE/ESD as an essential tool for development, the central policies have made EE more mechanical.
- While India is eagerly embracing ESD more quickly than EE, there is no clear understanding or articulation of the scope, processes and implication of ESD.
- Poor infrastructure facilities in schools, like drinking water, sanitation, number of classrooms, and teacher-student ratio, are hindering the implementation of EE.
- The EE strategies and processes lack region and target specificity which disconnect people and discourage them from participating in any EE processes.
- EE projects, programmes and activities are not adequately monitored and evaluated, which are important for their improvement.
- Documentation of the EE processes is not consistent which is leading to the loss of important information and knowledge generated, much needed for planning.
- Financial outlays for EE/ESD still remain at a sub-critical level. However, funds earmarked for EE in the Government are underutilized. This indicates a poor level of programme planning and management.
- Lack of coordination between Government departments working in the area of EE has resulted in repetition of programmes, wastage of funds and other resources. For instance, EE has been mainstreamed in schools, but the pre-service teacher training programmes, like the B.Ed., are still offering it as an optional subject.
- Strengths of community, youth in particular, NGOs, corporate sector and other academic institutions are often not recognized by the government and hence remain underutilized.

- Level of programmes and reporting in the mass media are insignificant due to various reasons including political pressure.
- Diverse biogeography; over 1.2 billion people; diverse cultures and languages (22 officially recognized languages each spoken by more than one million people except Sanskrit; around 33 other languages and 2000 dialects identified in India); poverty and low levels of literacy among women (around 65 % as we are nearing 65 years of Independence, census 2011); fluid political situations put pressure on implementation of EE.

6.13 Taking It from Here

"The volume of education … continues to increase, yet so do the pollution, exhaustion of resources, and dangers of ecological catastrophe. If still more education is to save us, it would have been a different kind: an education that takes us into depth of things" (Sterling and Schumacher 2001: 21). "EE in India has always been seen in the development context and therefore much of this can be seen as ESD. ESD needs to build on the foundation already laid and, over the next few years, become an integral part of how subjects are treated" (CEE 2009: 3). However, India also has to equip itself to fight new challenges in this endeavor. The way forward may be envisaged as a ten point approach.

1. India has to define clearly the scope of ESD, understand its processes and implications before it pushes ESD. The central role of ESD has to be clearly stated in all the state policies, programmes and project plans so that ESD is not brushed aside like EE as a supplement.
2. It is important to build strong evaluation mechanisms into all ESD activities in order to take stock of the outcomes and desired results and thus improve activities.
3. Environmental educators need to strengthen their capacities in ESD to be able to empower people to undertake environmental action. There is a need for educators to partner with academic and research institutions to improve the quality of training.
4. The success of ESD will depend upon the synergy between various departments within the Government and their key institutions. ESD programmes are not exclusive and cannot be planned and implemented in isolation.
5. There is a need to increase and improve partnerships with the Government institutions, NGOs, academic and research institutions and the corporate sector. The strengths and roles of each of the partners have to be understood in the given context. Such a partnership can make ESD more effective and take it to the next level.
6. Processes which will give people ownership of ESD programmes need to be thought of. This, in the long run, will help resolve conflicts related to natural resources and contribute to the process of sustainable development. Ensuring full community participation in ESD processes using various approaches is

necessary to make ESD a people's movement and to get the desired results of environmental conservation

7. Facilities, like zoos and botanical gardens, should be seen from the perspective of ESD. They offer myriad opportunities for spreading awareness among thousands of visitors. Small and big towns across India need to be upgraded, improved and strengthened in their existing facilities and new infrastructure should be created where necessary.
8. Local specific programmes and materials including training materials, like manuals, charts, simple kits (water testing and soil testing) for children, teachers, NGOs, policy makers, the corporate sector and the general public need to be brought out in the local languages to enrich the information, knowledge, skills and communication mechanisms.
9. Media need to play a proactive role with responsible reporting since it is an important vehicle to transfer information and knowledge to the masses.
10. India should harness the strengths of educational institutions since they serve as wonderful centres for data collection, pooling, storage and even environmental monitoring and information dissemination. Youth in particular, when trained can bring in novel and creative ideas to ESD processes.

With the new thinking emerging in this decade of ESD, there are several exciting opportunities coming in the way of India to redirect environmental consciousness and actions for its sustainable future (See Box 6.7).

Box 6.7: The Decade of Education for Sustainable Development (DESD)
The United Nations has promoted Education for Sustainable Development (ESD) in the decade starting from 2005 to 2014 through UNESCO, the lead agency in the promotion of ESD. Working with partners at all levels and from all spheres like government, civil society, NGOs and private sector UNESCO continues to strengthen and scale up ESD actions through Global Action Programme (GAP) beyond DESD.

India acknowledges ESD as a major driver of change for Sustainable Development. It is participating actively in the decade to promote ESD across the country. Indian vision of DESD is based on the commitment to sustainability rooted in the century old traditions of living in harmony with nature and its elements. It has hosted the first international conference of DESD at the beginning of the decade facilitated by CEE at Ahmedabad in Western India in January 2005. Several Government Ministries and Departments are putting information and awareness, capacity building in to their programmes.

India has been represented in both UNESCO reference group for DESD and the monitoring and evaluation group. The Indian national commission for UNESCO has appointed a high level committee to guide the implementation of DESD in India. (CEE 2009: 5). Through CEE India is also promoting GAP to further scale up ESD.

Growing concern about environmental issues, including climate change among people, rapid growth of mass media like radio and television, increasing availability of the internet facility, improvement in the satellite communication facility for cost effective training, introduction of a green agenda into Corporate Social Responsibility (CSR) plans by the industries and the corporate sector, shifting interests of consumers towards organic products, rising literacy rate, reducing school dropout rate, introduction of green benches (exclusive benches in the Supreme Court and the High Courts in India dedicated to address the cases related to environmental protection and conservation of natural resources) by the states to exclusively deal with environmental issues – are only some examples of the promising opportunities. India needs to utilize all these crucial opportunities, summon all her wisdom to bring in the change in Education for Sustainable Development.

References

Acharya, K. (2006). The water women: A case study of tank restoration. In S. Kothari (Ed.), *Seeds of hope*. New Delhi: Orient Longman.

Bankoti, T. S. (2008). *Chipko movement*. New Delhi: Global Vision Publishing House.

Bharucha, E. (2004). *Textbook for environmental studies*. UGC and BVIEER. http://www.ugc.ac.in/oldpdf/modelcurriculum/env.pdf. Accessed 12 May 2012.

BVIEER. (2002). *Study of status of infusion of environmental concepts in school curricula and the effectiveness of its delivery – A report*. Pune: Bharati Vidyapeeth Institute of Environment Education and Research (BVIEER).

CEE. (2002). *Storm over Silent Valley*. In Towards sustainability; stories from India. New Delhi: Ministry of Environment and Forests, Government of India.

CEE. (2009). *Towards a new development paradigm: Education for sustainable development*. India report to the world conference on ESD, Bonn, Germany 2009. Ahmedabad: Centre for Environment Education (CEE).

CEE. (2010). *Young in green action; inspiring stories from the national green corps*. Ministry of Environment and Forests, Government of India and Centre for Environment Education (CEE).

CEE. (2011). *National environment awareness campaign* 2011–2012. Ministry of Environment and Forests, Government of India and Centre for Environment Education (CEE).

Chico, M. http://www.uwcsea.edu.sg/page.cfm?p=1088. Accessed 2 Apr 2013.

Constitution of India, National Portal of India. http://india.gov.in/my-government/constitution-india/constitution-india-full-text. Accessed 12 May 2012.

Doran, P. (2002). *World summit on sustainable development (Johannesburg) – An assessment for IISD*. www.iisd.org/pdf/2002/wssd_assessment.pdf. Accessed on 25 Apr 2016.

DPEP. (1994). *District primary education programme*. www.educationforallinindia.com/page81.html. Accessed 25 Apr2012.

ENVIS Centres. http://www.envis.nic.in/centres.html. Accessed 10 Mar 2013.

Guha, R. (1989). *The unquiet woods: Ecological change and peasant resistance*. Berkeley: University of California Press.

Hegde, P. (2010). *The Appiko movement: Forest conservation in Southern India*. Cultural Survival. http://www.culturalsurvival.org/publications/cultural-survival-quarterly/india/appiko-movement-forest-conservation-southern-india. Accessed 8 Mar 2013.

INSA. (2001). *Pursuit and promotion of science: The Indian experience. Indian National Science Academy (INSA) New Delhi*. http://www.iisc.ernet.in/insa/ch33.pdf. Accessed 2 April 2013.

Machhiwalla, T. (undated). *Nali-Kali – A not so silent revolution for joyful learning, seeds of hope*. New Delhi: Lokayan. planningcommission.nic.in/reports/sereport/ser/seeds/seed_edu.pdf. Accessed 2 Apr 2013.

MHRD. (1998). *National policy of education 1986.* Department of Education, Ministry of Human Resource Development (MHRD), Government of India. http://mhrd.gov.in/sites/upload_files/mhrd/files/NPE86-mod92.pdf. Accessed 5 Mar 2013.

MoEF. (2002). Ministry of Environment and Forests, Government of India.http://envfor.nic.in/mef/intro.html 2002, http://moef.nic.in/modules/divisions/ee/?f=formal-ed. Accessed 10 Mar 2013.

MoEF. (2009). *Lapsing of the coastal management zone (CMZ) notification, 2008.* New Delhi, Ministry of Environment and Forests (MoEF), Government of India. http://envfor.nic.in/divisions/iass/2009-07-24%20Press%20Release%20-%20CMZ%20Lapse.pdf. Accessed 6 Mar 2012.

NCERT. (2005a). *National curriculum framework 2005.* New Delhi: National Council of Educational Research and Training (NCERT). http://www.ncert.nic.in/html/pdf/schoolcurriculum/framework05/prelims.pdf. Accessed 15 April 2012.

NCERT. (2005b). *National curriculum framework 2005.* New Delhi: National Council of Educational Research and Training (NCERT). www.ncert.nic.in/rightside/links/pdf/framework/nf2005.pdf. Accessed 15 Apr 2012.

Planning Commission, Government of India. (1970). *4th five year plan, 1969–1974.* http://planningcommission.nic.in/plans/planrel/fiveyr/welcome.html. Accessed 2 Apr 2013.

Planning Commission, Government of India. (1981). *6th five year plan, 1980–1985.* http://planningcommission.nic.in/plans/planrel/fiveyr/welcome.html. Accessed 2 Apr 2013.

Planning Commission, Government of India. (1985). *7th five year plan, 1985–1990, Vol 2.* http://planningcommission.nic.in/plans/planrel/fiveyr/welcome.html. Accessed 2 Apr 2013.

Planning Commission, Government of India. (2002). *India vision 2020.* http://planningcommission.nic.in/reports/genrep/pl_vsn2020.pdf. Accessed 2 Apr 2013.

Sarabhai, K. (2000, November 20–24). *Strategies in environmental education – Experiences from India.* In the international meeting of experts in environmental education. Santiago de Compostela, Spain. http://www.ceeindia.org/newintra/papers/Strategies%20in%20Environmental%20Education.htm. Accessed 2 April 2013.

Shiva, V. (1988). Women in the forest. In *Staying alive: Women, ecology and survival in India.* New Delhi: Zed Books.

Shiva, V., & Bandyopadhya, J. (1986). The evolution, structure and impact of the chipko movement. *Mountain Research and Development, 6*(2), 133–142.

Sinha, G. N. (2006). *British India's forestry and modern environmentalism.* http://www.sfri.org/images/general/article_brit_ind.pdf. Accessed 2 Apr 2013.

Sterling, S. R. (2001). *Sustainable education: Re-visioning learning and change* (Schumacher briefings 6). Bristol: Schumacher Society.

The Supreme Court of India. (2003). *Environmental awareness education – Supreme court directions dated 22.11.1991.* 2003 SOL Case No. 865. http://old.downtoearth.org.in/html/sc_directive.htm. Accessed 4 Apr 2013.

UNEP. (Not dated). *Declaration of the United Nations conference on the human environment.* http://www.uenp.org/documents.multilingual/default.asp?documentid=97&articleid=1503. Accessed on 25 Apr 2016.

UNESCO-UNEP. (1992). *Environmental education activities for primary schools, suggestions for making and using low-cost equipment.* UNESCO UNEP International Environmental Education Programme, Environmental Education Series 21 Produced by International Centre for Conservation Education.

Chapter 7
Developing Textbooks in India: Some Experiences of Introducing the ESD Perspective

Mamata Pandya

7.1 Introduction

The idea of education for sustainability is not new for India. It was the cornerstone of the Mahatma Gandhi's proposal for Nai Taleem, a spiritual principle which states that knowledge and work are not separate and which was used by Gandhi for an educational curriculum with the same name based on this educational principle. This New Education also known as Basic Education. This vision of a new social order was first shared in 1937, a decade before India became independent from British rule. For Gandhi, the goals of education and society were not separate. Nai Taleem was to achieve a harmonious development of head, heart and hand, based on sound moral principles and correlation with community (National Council of Rural Institutes 2009) (http://www.ncri.in/html/naitalim.html. Accessed on March 5, 2012). A key element of this vision of education was linking the curriculum with productive activity and the social environment. This was a radical idea not only in pedagogic terms but also in social terms.

The thinking underlying Basic Education was accepted as a policy guideline for education in the early years after India attained independence in 1947, but was somewhat abandoned in the early 1960s. Kumar (2004) states that the critical role of education as an important means of realising the vision of the Constitution of India was well recognised. The vision enunciated the construction of a liberal society based on democracy and social justice. In the more than six decades since India became independent, there have been a number of initiatives at the policy level, as well as through the formal and non-formal institutions towards achieving this vision. "The transformative role of education in relation to the social order and the values

M. Pandya (✉)
Independent Consultant, Educator and Editor,
Formerly Senior Programme Director, CEE, Ahmedabad, India
e-mail: mamata.pandya@gmail.com

R.N. Gorana, P.R. Kanaujia (eds.), *Reorienting Educational Efforts for Sustainable Development*, Schooling for Sustainable Development,
DOI 10.1007/978-94-017-7622-6_7

underpinning it has inspired considerable scholarship in recent decades but the answer has remained elusive and ambivalent (p. 15)". This chapter traces the chronology of the developments in the area of educational policy at the national level, with special reference to the quality and transformative role of education and its translation into national curriculum frameworks, syllabi and textbooks. It tracks parallel international developments in the theory and practice of quality education and its close linkages with Education for Sustainable Development (ESD). The discussion illustrates how India's National Curriculum Framework (NCF) 2005 demonstrates the process of infusing a 'Sustainability Perspective' (ESD as the present day's articulation of Basic Education and a self-sustaining society in a global and knowledge society) into the formal school system, in the context of the developmental and educational requirements of the country. The chapter shares the process of developing Environmental Studies (EVS) textbooks for classes III, IV, V and how this has provided the opportunity to infuse the intent and essence of both the NCF 2005 and ESD. It discusses some of the features of these textbooks that reflect the characteristics of ESD and quality education.

7.2 Educational Planning and Administrative Structure

India is a federation of 28 States and seven Union Territories with two tiers of Government, one at the level of the Union (known as Central Government), and the other at the federal level (known as State Government). As a federal system, there are well defined powers of the Central and State Governments. The Indian Constitution includes three lists enumerating the powers of the Central and the State Governments: the Union, State and Concurrent lists.

Immediately after the country attained independence from British rule in 1947, providing education became the responsibility of the Indian government in keeping with its welfarist function, (Rao 1999: 3). Following the enactment of the Indian Constitution in 1950, education became primarily the responsibility of the State Governments. The Central Government could only provide guidance to the states on education-related issues. In 1976, with the enactment of the 42nd Constitutional Amendment, education was transferred to the Concurrent List, with the objective to promote meaningful educational partnerships between the Central and State Governments. This required that both the Central and State Governments were to have a joint responsibility in the area of education.

The Department of Education of the Ministry of Human Resource Development (MHRD) at the Central level shares with the States the responsibility for educational planning. The MHRD, however, is not the sole ministry that is concerned with education. Several ministries, including the Ministries of Environment and Forests; Agriculture; Health; and Law also engage with large education related programmes. The Central Government establishes broad education policies for the development of school curricula and management practices that serve as guidelines for states. Decisions regarding the organization, structure of education, curriculum and textbook development are largely the concern of the states. Within each state there is a State Education Department with academic and administrative structures. The efforts of the Central Government in educational planning and administration is supplemented by the efforts and inputs of other expert bodies some of which are described here.

The National Council of Educational Research and Training (NCERT), constituted in 1961, plays a significant role in curriculum development and in the training of teachers. It guides the states by providing a national framework for curriculum development, and building capacities for all stages of school education. The Central Advisory Board of Education (CABE) is the highest advisory body to advise the Central and State Governments in the field of education. The National Council for Teacher Education (NCTE), formed in 1973, advises the Central and State Governments on all matters pertaining to teacher education. The National University of Educational Planning and Administration (NUEPA), looks after capacity building and research in planning and management of education in India as well as in South Asia.

Within the country, there are parallel systems of school education. While the majority of schools are affiliated to their respective State Board of Education, private schools are also affiliated to the State Boards or Central Board of Secondary Education (CBSE), or Indian School Certificate (ICS) systems, and the International Baccalaureate (IB) each with its own examination board and pattern, syllabus and textbooks.

There are three main school types: government, aided, and private. Schools run by the Central, State or local governments are referred to as 'government' schools. Schools run by private managements, but funded largely by government aid, are known as private aided or just 'aided' schools. Schools funded in whole or in part by charging their students tuition, rather than relying on government funding, are known as private schools, and these schools retain the right to select their students. The multiplicity of Education Boards and examination systems has led to considerable diversity in syllabi and textbooks in the country.

7.3 Some Milestones in the Development of Education Policy

During the post-independence period, education was recognized as a factor vital to national development. There was a great deal of thinking and deliberation about what should be the goals and objectives of education. National education policies have evolved through a process of extensive consultations in which all the states and Union Territories actively participate. Periodically, the Central/State Governments appoint commissions and committees to examine various aspects of education. In addition, countrywide debates take place on various educational issues. The recommendations of various commissions, committees and national seminars, and the consensus that emerges during the national debates form the basis for India's education policies.

Mahatma Gandhi had visualized education as a means of transforming a society that was characterized by conflicts, injustice and inequality, and emphasized the self-reliance and dignity of the individual. The early Commissions set up after Independence (The Secondary Education Commission 1952–1953) and the Education Commission (1964–1966) articulated the themes emerging from Gandhi's educational philosophy, in the changed socio-political context and the new nation's development priorities. The Education Commission was appointed in 1964 under the chairmanship of D.S. Kothari. The Kothari Commission Report titled *Education and National Development* clearly reflected the vision of education as the most powerful instrument of national development. It perceived a broader role for education in enriching the quality of life. It asserted that besides knowledge and skills, education should also be concerned with the 'inner content' of people's lives, with ideas and idealism, and strengthening of the spirit. "We need a balance between three overlapping divisions of education (at all levels) which may perhaps be described as: tactical, strategic and humanistic. The first refers to theoretical and practical knowledge of life-long utility, the second to knowledge of life-long utility and value, and the third relates to quality and meaning of life"

(Daulat Singh Kothari, The Architect of Defence Science in India.) (http://www.vigyanprasar.gov.in/scientists/dkothari.htm. Accessed on February 15, 2012). The Report was hailed as a landmark in educational sectors in India and other developing countries.

Kothari's deep concern for education led to his association with NCERT since its inception. He not only conceptualised its role and functions but also gave a blueprint for its future development in the Report of the Education Commission (1964–1966). J.P. Naik who was member-secretary of the Education Commission, also made valiant efforts to get some of the commission's more important recommendations accepted and enforced by the government. As Educational Advisor in the central Ministry of Education, he helped to establish several new educational institutions, such as NCERT, the National Institute of Educational Planning and Administration Educational Planning and Administration (NIEPA), now NUEPA.

The vision of the Kothari Commission played a significant role in the formulation of the National Policy on Education (NPE) 1968. This marked a significant step in the history of education in post-independence India (Rao 1999: 44):

> The NPE 1968 aimed to promote national progress, a sense of common citizenship and culture, and to strengthen national integration. It laid stress on the need for a radical reconstruction of the education system to improve its quality at all stages, and give much greater attention to science and technology, the cultivation of moral values and a close relation between education and the life of the people (MHRD 2011: 1) (http://education.nic.in/NatPol.asp. Accessed on February 22, 2012).

Following the inclusion of education in the Concurrent List in 1976, renewed priority was assigned to education in order to meet the new challenges and social needs of the country. There were extensive deliberations by various national committees on the country's education system and policy which culminated with the decision to evolve a National Curricular Framework which would indicate the directions in which the educational system of the country was to proceed in order to implement its education policy. This took the form of the NPE 1986. This was the first time that the country as a whole had a uniform national policy on education. It proposed a national framework for the curriculum as a means of evolving a national system of education capable of responding to India's diversity of geographical and cultural milieu, while ensuring a common core of values along with academic components (NCERT 2005: 4). It included a specific mention of the need to create awareness of environmental concerns through infusing environmental components in the syllabi at all levels of education.

NCERT was entrusted with the responsibility of developing the National Curricular Framework (NCF) and reviewing the framework at regular intervals in the light of emergent trends in education. In this role, it carried out several studies and consultations culminating in the *National Curriculum Framework for Elementary and Secondary Education 1988*. This was the first document detailing a national curricular framework in schools. The curriculum was designed to enable the learner to acquire knowledge to develop concepts and inculcate values commensurate with the social, cultural, economic and environmental realities at the national and international levels (NCERT 2000: 19).

Despite the stated intent and comprehensive approach of these national documents towards creating a nurturing environment for the development of competencies and values, school education in the 1990s, at all stages, came to be driven by examinations, high-levels of competition, and information-loaded textbooks. In response to the great concern about the increasing academic burden on students and the unsatisfactory quality of learning, the MHRD set up a National Advisory Committee in March 1992 under the chairmanship of Professor Yash Pal. The committee undertook a nationwide consultative process which provided the inputs to its report titled *Learning Without Burden* published in 1993.

The recommendations of this report span the entire educational process including the process of curriculum development, textbook writing and teacher preparation, with the aim of bringing a sense of "joy" of being involved in the educational process, both to teachers as well as the students. It is in this context that the NCERT initiated the process of developing the next curricular framework for school education. This was titled the *National Curriculum Framework for School Education* (NCFSE) 2000. This document recommended the need for "paradigm shifts to support a curriculum that values the interaction of the process and the content" (NCERT 2000: 39–40). It emphasized that meaningful school curriculum has to be responsive to society, reflecting the needs and aspirations of its learners (NCERT 2000: 8) and suggested that the curriculum must stand on the three pillars of relevance, equity and excellence. The 2000 Framework emphasized that the school curriculum should aim at "enabling learners to acquire knowledge, develop understanding and inculcate skills, positive attitudes, values and habits conducive to the all-round development of their personality" (NCERT 2000: 39). Further, it stated that education should promote the quality of learning to listen to others, to learn from the events that surround us, and understand the economic, social and political environment, whether at a national or global level (NCERT 2000: 15).

While the NCFSE 2000 laid emphasis on addressing the issues of curricular load and an examination-oriented system, these continued to be the dominant characteristics of school education in India, even in the new decade. In 2004, MHRD asked NCERT to review the NCFSE 2000, especially with reference to the above concerns. To facilitate this process a National Steering Committee and 21 National Focus Groups were set up to cover pertinent areas relevant for curricular redesign. The Committee and the Focus Groups steered a nationwide process of intensive deliberations which saw the participation of a wide range of stakeholders, from scholars to parents and students, and input from numerous government as well as non-governmental institutions. The outcome was a document titled *National Curriculum Framework (NCF) 2005*.

The NCF 2005 was a response to the new developments and concerns facing the country. It attempted to address what Naik (1975) described as the 'elusive triangle' of Indian education – equality, quantity and quality. In terms of equality and quantity, NCF 2005 reflected a commitment to ensuring that all children have access to education, irrespective of socio-economic backgrounds and variations in physical and intellectual characteristics. It cautioned that universalization of elementary education would be meaningless unless the quality aspect was seriously addressed.

The NCF 2005 envisioned the transformation of the Indian education system. It emphasized 'learning without burden' and 'child-centred' education. It recommended systemic changes as markers of curricular reform. The NCF 2005, therefore, has many more facets than are obviously visible or explicit. It is what may be described as the 'intended' curriculum in that it is shaped by the general ideology (or, as it is often called, philosophy) of education. This intended curriculum provided the guidelines for developing the written formal curriculum, which would become the officially approved plan for instruction to be implemented by teachers as the 'active' curriculum as stated by Pingel (2009: 28) in his chapter concerned with the pedagogical environment. The process of translating the intended curriculum (NCERT 2005) into the active curriculum (the syllabus and textbooks) was influenced directly, as well as in spirit, by thinking at the international level.

7.4 Quality and Scope of Education: The International Context

Interestingly, a similar chronology and parallel concerns regarding the quality and scope of education were being expressed globally during the period between 1990 and 2005. In November 1991 the General Conference of UNESCO invited the Director General "to convene an international commission to reflect on education and learning for the 21st century". In response, the International Commission on Education for the twenty-first century was set up in 1993. The Commission comprised a group of fifteen eminent figures from across the world, and was chaired by Jacques Delors. The report of the Commission was submitted to UNESCO and published in 1996 under the title *Learning: The Treasure Within.* This report underlined the fundamental role of education in the attempt to "attain the ideals of peace, freedom and social justice". It emphasized the importance of education in both personal as well as social development, and saw it as one of the principal means available "to foster a deeper and more harmonious form of human development, and thereby to reduce poverty, exclusion, ignorance, oppression and war"(Delors 1996: 11). The report further emphasized that it was vital to conceive education in a more encompassing fashion and that such a vision should inform and guide future educational reforms and policy, in relation both to contents and methods. It proposed education throughout life based on four pillars: "Learning to know, Learning to do; Learning to live together, and Learning to be" (Delors 1996: 37).

In India, the NCF 2000 included a reference to this report, and reflected a similar vision when it stated that "the curriculum must meet the learner's needs, societal expectation, community aspirations and international comparisons" (NCERT 2005: 7). The Delors Commission's Report also reflected the growing understanding that economic development alone might not be sufficient as the basis for an equitable and just world and that environment and development could not be seen as separate challenges; rather, that they were linked in a complex system of cause and effect. This approach was clearly being reflected in the international conferences and deliberations on environment and development through the 1990s. Sustainable development was included and described as "development that meets the need of the present without compromising the ability of future generations to meet their own needs" as stated by the World Commission on Environment and Development (WCED 1987: 9). This was also the period when there was a focus on the formulation and development of the vision and framework for the reorientation of education towards the challenges and demands of sustainable development (SD). The need for education to be reoriented towards sustainable development was clearly articulated in Agenda 21, Chapter 36: Promoting Education, Public Awareness and Training, the outcome of the United Nations Conference on Environment and Development held at Rio de Janeiro in 1992. It was recognized and agreed, that "education is critical for achieving environmental and ethical awareness, values and attitudes, skills and behavior consistent with sustainable development and for effective public participation in decision-making. Both formal and non-formal education is indispensable to changing people's attitudes, so that they have the capacity to assess and address their SD concerns" (36.3) (http://www.gdrc.org/sustdev/un-desd/c36-a21.html. Accessed on March 5, 2012).

This recognition and articulation set the foundations for the subsequent developments of education for environment and sustainability. Scoullos and Malotidi (2004) highlighted this wherein education was seen not as an end in itself but as an indispensable instrument for achieving a sustainable future. The scope of education broadened, touching on and integrating the notions of population, poverty, environmental degradation, democracy, human rights and peace, development and interdependence. The outcomes of the World Summit on Sustainable Development held in Johannesburg in 2002 underpinned the need to integrate sustainable development into education systems at all levels of education, in order to promote education as a 'key agent for change'. The global focus on the critical role of education in achieving SD was reflected in the declaration of the United Nations Decade of Education for Sustainable Development (UN DESD) from 2005 to 2014. *The International Implementation Strategy for the Decade* described the declaration of DESD as an attempt to integrate the principles, values, and practices of sustainable development into all aspects of education and learning, and to reorient educational programmes, policies and practices so that education plays its part in building the capacities of all members of society to work together for a sustainable future. The strategy reaffirms that education and communication are the key drivers of change towards sustainable development. The International Implementation Scheme reiterated that, "It [the DESD] sees education as an enabling factor to help understand ourselves and others, and our links with the wider natural and social environment and this understanding serves as a durable basis for building respect. Along with a sense of responsibility, respect, exploration and dialogue, it visualises that Education for Sustainable Development (ESD) would aim to move us to adopting behaviors and practices which enable all to live a full life without being deprived of basics" (UNESCO 2005a: 6).

In this context, ESD shares the characteristics of any high quality learning experience with the additional criterion that the process of learning/teaching must model the values of sustainable development itself. In India, NCF 2005 reflected this global thinking when it described quality education as being inclusive of a concern for quality of life in all its dimensions. It stressed that this was the reason that a concern for peace, protection of the environment, and a predisposition towards social change must be viewed as core components of quality and not merely as value premises (NCERT 2005: 9).

7.5 Curriculum Development

The process of curriculum development in India lies between the two extremes of centralization and decentralization. As already discussed, the Central Government, from time to time, reviews and formulates the national policy on education which provides broad guidelines with respect to the content and process of education for different stages. These guidelines are further elaborated by NCERT in the form of the curriculum framework. This framework, prepared at the Central level, provides a broad overview of the school curriculum, including general objectives, subject-wise objectives, suggested schemes of study, and guidelines for the transaction of the curriculum and the evaluation of pupil outcomes. The NCERT curriculum framework is suggestive rather than prescriptive and it is not enforceable by law in the states.

The syllabi and instructional materials developed by NCERT are used in the schools run by Central organizations across the country. The states consider whether to adopt or adapt the NCERT syllabi and instructional materials. However, these are by and large accepted by many of the states because of NCERT's credibility, and the participatory development approach that it follows. At this point, it would be pertinent to review how the word 'curriculum' is construed. The Focus Group on Curriculum, Syllabus and Textbooks observed that there is a tendency to take too wide a definition of curriculum as stated in the position paper (NCERT 2006a: 11):

> Discussions in curriculum seem to say too often that everything that happens in the school is part of the curriculum. On the other hand, at practical level, syllabus, construed as a list of objectives and topics in a particular subject, is often referred to as curriculum. These two tendencies may look contradictory in the first glance – literature claims everything to be curriculum and the planned/provisioned view covers an extremely narrow part of the school experience – but the very declaration that everything is curriculum is a challenge to planning. Therefore, what is considered essential is planned, and rest of the 'broad vision' of curriculum is left to happening by chance.

The NCF 2005 started by revisiting an understanding of the role and scope of curriculum, with reference to the four fundamental questions for developing any curriculum and plan of instruction as outlined by Tyler:

- What educational purposes should the schools seek to attain?
- What educational experiences can be provided, that are likely to attain these purposes?
- How can these educational experiences be effectively organised?
- How can we determine whether these purposes are being attained? (Tyler, 1949 as quoted in the NCERT 2006a: 11)

In addressing these, NCF 2005 envisioned that the aims of education should simultaneously reflect the current needs and aspirations of a society as well as its lasting values and the immediate concerns of a community as well as broad human concerns (NCERT 2005: 10). The document suggested some guiding principles for curriculum development towards achieving these aims of education:

- Connecting knowledge to life outside the school;
- Ensuring that the learning is shifted away from rote methods;
- Enriching the curriculum to provide for overall development of children rather than remain textbook centric;
- Making examinations more flexible and integrated into classroom life;
- Nurturing an overriding identity informed by caring concerns within the democratic polity of the country (p.viii).

The application of these principles would in a sense help to achieve the broader definition of curriculum "as that set of planned activities which are designed to implement a particular educational aim – set of such aims – in terms of the content of what is to be taught and the knowledge, skills and attitudes which are to be deliberately fostered together with statements of criteria for selection of content, and choices in methods, materials and evaluation"(NCERT 2006a: 19). This was the background to the process, initiated by NCERT in 2005, of development of the syllabus and textbooks within the National Curriculum Framework for all subjects for all the levels of the school system. This chapter, as already mentioned, focuses on the EVS textbooks for classes III, IV and V (students aged 8 years, 9 years and 10 years respectively) developed as part of this process. In the context, it is important to understand how the area of environmental education has been perceived and situated in the educational policies and practice in India.

7.6 Teaching and Learning About the Environment in the Curriculum

The importance of creating environmental consciousness across all age groups and sections of society was clearly articulated in the NPE 1986:

> There is a paramount need to create a consciousness of the environment. It must permeate all ages and all sections of society, beginning with the child. Environmental consciousness should inform teaching in schools and colleges. This aspect will be integrated in the entire education process (MHRD 2011: 23) (http://education.nic.in/NatPol.asp. Accessed on February 22, 2012).

Education in schools in India is structured along a '10+2' pattern. Ten years of general education are followed by 2 years of senior secondary education with disciplinary specialization in preparation for university education. The general tier of school education is further divided into the primary (Classes I–IV or V), upper primary or middle (Classes V or VI–VIII) and secondary (Classes IX–X) stages. Three Curriculum Frameworks (1988, 2000 and 2005), subsequent to NPE 1986, reiterated the importance of environmental education (EE). The approach adopted by NCERT has been to offer environment as a composite subject at primary level. At middle and secondary school levels, environmental concepts, issues and concerns are infused into the science, social science and language syllabi. At the higher secondary (+2) stage, concepts related to environment and development have been integrated into the syllabi of biology, chemistry, physics, geography, economics, sociology and political science. In 1991, a significant and far-reaching event that pushed environmental education more clearly into the formal education system in India was the Directive of the country's Supreme Court in response to a Public Interest Litigation. It emphasised the need to make Indians more environmentally sensitive and responsible citizens. In its order the court directed that:

> We would require every State Government and every Education Board connected with education up to the matriculation stage [*and beyond*]…to immediately take steps to enforce compulsory education on environment in a graded way (http://www.greenteacher.org/?page_id=69. Accessed 19th February, 2012).

In response to the Supreme Court's directive and the requirements of the NPE 1986, several steps have been taken by government agencies and NGOs to introduce and promote environmental education. The NCF 2005 Syllabus for Classes at the Elementary Level re-endorsed the recommendation of NCF 2000 that EVS be taught as an integrated course for the entire primary stage and called for further strengthening of the integrated approach for EVS during the primary years. The National Focus Group on Curriculum, Syllabus, and Textbooks, also opined that at the primary level there may not be sufficient "conceptual basis for any clear demarcation of sciences and social sciences but it could be possible to introduce ways of looking at the natural and social world in the form of activities, ways of data collection, and making sense out of them" (NCERT 2006a: 31).

The main focus of EVS as spelled out in the Focus Group Paper on Habitat and Learning, 2006 is:

> to expose students to the real-life world, natural and social, in which they live; to enable them to analyze, evaluate and draw inferences about problems and concerns related to the environment; to add, where possible, to our understanding of environmental issues; and to promote positive environmental actions in order to facilitate the move towards sustainable development (NCERT 2006b: 4).

The NCF 2005 also indicated some of the objectives of EVS at the primary stage:

- To guide children to locate and comprehend relationships between the natural, social and cultural environment;
- To develop an understanding based on observation and illustration, drawn from lived experiences and physical, biological, social and cultural aspects of life, rather than observations; and
- To create cognitive capacity and resourcefulness to make the child curious about social phenomena, starting with the family and moving on to wider spaces (NCERT 2006c: 90).

The syllabus for EVS at the classes III, IV and V was designed to achieve these objectives. The intent, approach and content of the syllabus reflect the spirit of the NCF 2005 as well as some of the characteristic principles of ESD as described in the UN DESD International Implementation Scheme.

7.7 From Curriculum to Syllabus in EVS

The National Focus Group Report on Curriculum, Syllabus and Textbooks expressed one of the serious concerns in the Indian educational system which has been that "practices adopted for development of curricula, syllabus and textbooks are largely guided by the patterns and requirements of the examination system rather than a mix of criteria based on the child's learning requirements, aims of education, and the socio-economic and cultural contexts of learners" (NCERT 2006a: 1).

Learning Without Burden, the Yash Pal Committee Report (MHRD 1993) mentioned earlier in the chapter, had raised the above concern by pointing out how the syllabus and textbooks gave an impression that the experts involved in preparing these had little knowledge of school and classroom realities. The report had expressed apprehension that this limitation could extend to the experts' possible ignorance of children and of the processes that children use for learning new ideas (MHRD 1993: 18). It further stated that the "lack of adequate opportunities for teachers to participate in the process of syllabus and textbook preparation was a major factor indirectly responsible for the problem of unrealistic syllabi or curriculum load" (MHRD 1993: 20). It recommended that the culture of writing textbooks be changed so as to involve a much larger number of teachers in their preparation. International thinking supports this recommendation. The *UNESCO Guidebook on Textbook Research and Textbook Revision* reiterates that in textbook development processes it is of the utmost importance that subject-oriented expert knowledge and educational experience must be given equal status and that both academic experts and teachers are able and willing to cooperate in the preparation of textbooks (Pingel 2009). To address these concerns, and to reflect the spirit of the NCF, the committees for the development of syllabus and textbooks involved a number of teachers and representatives of groups working closely with children, in formal as well as non-formal settings, in addition to subject experts and NCERT faculty.

7.8 The Syllabus

An appropriate prioritization of learning objectives and effective use of available time and space are important requirements for successful teaching. Equally important is the consideration and recognition of the learner's needs, cognitive style, gaps, aptitudes and motivation. All these need to be addressed at the syllabus planning stage itself. The NCERT syllabus attempted to address these concerns and requirements with respect to scope, content, and methodology, including the investigation of curricular structures based on trans-disciplinary approaches, using thematic modules instead of isolated disciplines.

School curricula are usually organised around various subjects. The subjects themselves have their bases in traditions and disciplinary knowledge. Though the subject-based organisation of curriculum is largely accepted, it is criticised for putting knowledge in watertight compartments and, thereby, fragmentising it. The Focus Group on Curriculum, Syllabus, and Textbooks expressed the concern that this 'fragmentisation' of knowledge is said to be alien to the child's way of looking at the world and, therefore, unsuitable for developing a proper understanding of the world in which the child lives (NCERT 2006a: 24).

A significant feature of the 2005 syllabus for EVS is that it attempts to move away from isolated 'topics'. It proposes instead 'themes' which allow for a connected and inter-related understanding of different aspects of different environments – natural, social and cultural – to develop (see Table 7.1 A sample of NCERT's Environmental Studies Syllabus for Class IV). Going by this approach:

- The syllabus is organized not as a list of 'topics' but as 'themes'.
- The thematic approach helps to bring in perspectives from different subjects and disciplines.
- Each theme has sub-themes organized in a spiral and progressive manner.
- The sub-themes allow development of a connected and inter-related understanding.
- The themes in the syllabus do not begin by listing key concepts but by posing key questions which allow each child to think, apply, and develop his/her own understanding and articulate his or her own learnings and ideas.
- The activities are only suggestions and can be easily adapted to suit the local situations.

Table 7.1 A sample of NCERT's Environmental Studies syllabus for Class IV (NCERT 2006c: 115)

Questions	Key concepts/issues	Suggested resources	Suggested activities
Whom do trees belong to?	Neighbourhood and its plants; wild and domestic plants	Local knowledge, information about domestic and wild plants (National Book Trust books)	Listing of some common trees in the neighbourhood; discussion about ownership of trees; fruits that are not eaten by us
Which plants/trees around you are looked after by people – by whom? Which are not? Whom do they belong to? Who eats the fruit of trees that grow wild?	Fruits eaten by people living in forests		
	Cutting trees		
FOOD	From field to *mandi* – from market to house; grown by farmers; fruit trees, vegetables, cereals, pulses, oil seeds; Spices	Discussion with a vegetable seller/ retailer in the mandi/truck driver who transports food items	Listing plants children know that provide them food; bringing samples; common spices, observing and drawing samples, recognizing them by smell and taste
How we get our food			
How does food reach us? Who grows it? How you seen vegetables and fruits growing? How you seen plants of rice/ wheat/dal etc? What spices do you know? Which spices can we recognize by smelling or tasting			

The EVS syllabus for Class III to V is woven around the six interconnected themes. These themes provide opportunities for introducing aspects of science, social science and environmental education. The six themes are:

1. Family and Friends

    1.1 Relationships
    1.2 Work and Plan
    1.3 Animals
    1.4 Plants

2. Food
3. Shelter

4. Water
5. Travel
6. Things We Make and Do (p. 92)

The content of the themes is derived from the child's experiences rather than a prescriptive body of knowledge or subject. These are presented in a manner in which a child encounters them in daily life. Starting with these personal experiences, the child is enabled to "connect", and thereby link theoretical knowledge to his/her own life. The same themes are carried through the 3 years, moving outwards from the immediate 'self' of the child to include her family, the neighbourhood, the locality, and the community. Thus the child begins with an exploration of the familiar and expands her horizons as she grows. At the same time the child is also able to locate herself in different contexts – as a family member, a member of the school community and as a future citizen of the town/city and country. It is this syllabus that has provided the framework for the development of the EVS textbooks for classes III, IV and V. The author was part of the team that developed Environmental Studies (EVS) textbook for classes III, IV, and V. The process of textbook development spread across 3 years from 2005 to 2008.

7.9 Textbooks in the Indian School System

Textbooks are developed and published by NCERT and SCERTs (State Council of Educational Research and Training), as well as by private publishers. Textbooks developed by NCERT/SCERTS are distributed free of cost in government schools. Private schools are free to prescribe the textbooks to be used. At the primary level most of these are from private publishers. NCERT textbooks are directly used in some states and translated and adapted by many others. Kumar (2004: 24) observed how textbooks dominate the curriculum in the ordinary Indian school and how the teacher is bound by the textbook. Textbooks, in the Indian school system, are indispensable, both for teachers and students. Textbooks are often the one and only resource that both the teacher and the students possess. Each child has a set of textbooks for all the subjects for the class in which he/she studies. The prescriptive nature of the textbooks is evidence of how everything that happens in the school or classroom is based on the textbook itself. Textbooks inform the day-to-day organization of time in the classroom, and what and how much is learnt by the student to meet the requirements of the examination system. Kumar (2004, p. 25) observed how the textbook is a "structuring device, offering a programme of sequenced action" The teacher teaches what is written in the textbook, lesson by lesson, and limits recapitulation and evaluation to what is given as exercises at the end of each lesson.

The stature of the textbook and its pivotal role in the school system and in education throws up numerous challenges in itself. In addition, the process for developing the textbooks within the framework and guidelines of the NCF 2005 needed to address several other concerns:

How to write at a national level and yet reflect the multiple dimensions of schools and classrooms across the country?

How to address the tremendous diversity in respect of ecological, social, cultural, economic and religious dimensions?

How to organize and present the content in a perspective that views the environment as a totality, and how to avoid 'compartmentalizing' it through 'subjects' like science and social science.

How to provide an enabling common framework, with adequate in-built space for adaptation, right down to the level of the individual teacher?

How to design the textbook contents so as to give each the child opportunities to articulate his/her experiences and to develop logical arguments and skills, all with the existing realities and constraints of the classroom and teacher capacity?

How to provide evaluation tools that assess beyond knowledge gained by rote learning?

The challenge was also how to give form and life to the vision of the NCF 2005 which emphasized that the curriculum needs to be conceptualized as a structure that enables children to:

- Find their voices;
- Nurture their curiosity;
- To do things;
- Ask questions;
- Pursue investigations; and
- Share and integrate their experiences with school knowledge rather than test their ability to reproduce textual information.

7.10 The Textbook Development Process

In the lesson-to-lesson reality of the textbook, the writers and designers are faced with three challenges which are opportunities at the same time – how to synthesise the intent of the syllabus, the context of the learner, and the use of the textbook in actual classroom transactions. Textbooks need to be open to varied interpretation by the learner and the teacher, and be flexible enough to be reconstructed in the classroom. These challenges were addressed by the textbook development team which comprised school teachers, teacher educators, and practitioners and which came from diverse backgrounds and experiences. The development process for each book spanned a period of 8–10 months. Each year the process started with intensive brainstorming by the entire team to identify the format and key coverage of each thematic area for the relevant class (e.g. III, IV or V). Thereafter, smaller teams of two or three persons worked separately on drafting lesson/s based on their selected theme. These drafts were discussed in detail by the entire team at periodic intervals and suggestions/inputs noted to be included in the reworking by small teams. Once all of the draft chapters were developed, the entire team once again worked collectively at fine-tuning language, exercises and questions. This process of textbook development threw up a number of challenges as well as the opportunities to explore innovative ways to transform challenges into meaningful teaching-learning material.

7.11 The EVS Textbooks

The EVS textbooks for classes III, IV and V were published under the title *Looking Around*, in 2006, 2007, and 2008 respectively by NCERT. This title indicates the perspective that defines the process of teaching and learning about the environment. It assumes that the surroundings provide scope for observing, experiencing and connecting. It communicates the important pedagogical principle that the world around us and our interaction with it is an endless, rich resource for learning. It underlines that we can learn much by interacting with the world around us. The textbooks are an attempt to reiterate the quality of a good textbook as described by the National Focus Group on Curriculum, Syllabus, and Textbooks: "Any good textbook should lead the child to interact with the environment, peers, other people, etc., rather than be self-contained. It should function as a guide to construct understanding through active engagement with text, ideas, things, environment, and people rather than 'transferring knowledge as a finished product' (NCERT 2006a: 38).

The title of the EVS textbooks *Looking Around* is also indicative of the content, approach and key features of the textbooks that come closer to the student's own context and accommodate a broader base of information as opposed to that reflecting a specific milieu. It supports Sandell's view that, if education is to be beneficial to students, it is important that both the choice of content and the teaching approach take student preferences into consideration. It is also important that these choices are made in relation to the students' previous knowledge and experience, as well as allowing them opportunities to draw on these resources. If not, there is a risk that what is learnt during lessons will only be valid in the classroom and of no relevance for the students' life and actions outside school (Sandell et al. 2003: 232). In the *Looking Around* textbook series, the selection and organization of content, the treatment of the themes and the approach and methodologies come together to provide each child the space to become an active participant in looking around themselves, exploring and discovering.

The EVS textbooks reflect some of the key characteristics of ESD as described in the UN DESD International Implementation Scheme document:

- Inter-disciplinary and holistic: learning for sustainable development embedded in the whole curriculum, not as a separate subject;

- Values-driven: sharing the values and principles underpinning sustainable development;
- Critical thinking and problem solving: leading to confidence in addressing the dilemmas and challenges of sustainable development;
- Participatory decision-making: learners participate in decisions on how they are to learn;
- Locally relevant: addressing local as well as global issues, and using the language(s) which learners most commonly use.
- Multi-method: word, art, drama, debate, experience…different pedagogies which model the processes (UNESCO 2005b: 6).

The approach to the development of the EVS textbooks reflected "the option of introducing the idea of sustainable development, not as a new subject, but rather as an integrated perspective which is applied to the existing content in all subjects" (Sandell et al. 2003: 198). The content and its treatment are further discussed with reference to the EVS textbooks for classes III, IV and V.

7.12 Interdisciplinary and Holistic

A young child's world is not organized into compartments, but is an integrated experience of interacting with, and making sense of the world around. While interpreting the new curriculum and the new textbooks the key concern was to present the environment as a totality and avoid treating different aspects as in compartmentalised subjects such as 'science' and 'social science'. Just as the title of the books has moved away from the 'subject' label, the organization of content of the books also moves away from the traditional 'topics' to an exploration of dimensions of different disciplines through various themes. The approach was to achieve an integration of the different aspects of a theme and provide space for each child to explore and discover the world around. Therefore, concepts that are typically dealt with in biology such as plants, or animals that are studied in zoology, or which are at the primary stage framed in the categories of 'living and non-living', are introduced not as categories but through the child's familiar experiences and surroundings.

Besides thematic integration that synergises different subjects, integration is also facilitated through the different approaches and methods of transaction, assessment, and the 'values' being promoted. This supports the belief that when watertight compartments between subjects are broken, this can enrich not just the 'content' but also the process and outcomes of teaching and learning. A similar approach is supported by Sandell where working with sustainable development is a case of putting things into context and where the different areas and levels are seen as integral parts of a whole (Sandell et al. 2003: 199).

7.13 Values and Sensitivity

The UN DESD International Implementation Scheme stressed that understanding values is an essential part of understanding an individual's own world view and that of other peoples'. Understanding your own values, the values of the society you live in, and the values of others around the world is a central part of educating for a sustainable future. Each nation, cultural group and individual must learn the skills of recognizing its own values and assessing these values in the context of sustainability (UNESCO 2005b).

ESD is intrinsically about values, with respect at the centre. This involves respect for others, including those of present and future generations, for difference and diversity, for the environment and for the resources of the planet. The ways countries decide how to approach sustainable development will be closely linked to the values held in these societies for it is these values that define how personal decisions are made and how national legislation is written. Textbooks play an important role in communicating values, explicitly and implicitly. This role has been a continuing topic of debate. As Slater in Pingel pondered, "Do history, geography and social studies textbooks seek only to reflect society, or to change it? Do they seek to guarantee certain attitudes and values? Or do they more modestly seek to enable young people, with a foundation of knowledge, skills and insights, to make their own independent choices between alternative attitudes and values?" (Pingel 2009: 62).

Traditionally, textbooks in India, especially for EVS and social science tend to portray the 'ideal' or 'positive' by portraying 'role models' or advocating 'good behaviour'. Generally, the trend has been to ignore awkward, or unpleasant, areas or touch upon issues of inequity, discrimination and contradictions with platitudes like "We should respect our elders" or "Everyone is equal". In the last decade or so multi-perspectival methodologies and skill-oriented approaches that do not prescribe a fixed body of content issues are being implemented in more and more countries. As described by Pingel (2009: 40) in a chapter on multiculturalism in the *UNESCO Guidebook*, "Textbooks employ a narrative structure that integrates all components of the society and regards the mixture and growing together as a value in itself. They follow an integrative and inclusive concept rather than a separatist and exclusive model of collective identities."

The EVS textbooks attempted to follow this approach. In these textbooks, issues of equity and access to resources are sensitively raised in lessons that project these through a child's perspective. Children are encouraged to share their personal experiences and opinions through the in-text questions and exercises. There is an attempt to reinforce continually the fact that there are differences in the way people live and, while we may not do exactly what or how others do, it does not mean that those who are 'different' are 'bad'. It is important to sensitize children to the fact that difference is not bad.

The stories of physical, socio-economic, and cultural differences are told through the voices of children and often reflect the child's perception of the situation. The narratives invite children to understand such issues by questioning, analyzing, debating and relating with their everyday life and experiences rather than 'telling' the child what to think. The intent of exposing children to real issues through an engaging methodology is that it will nurture and shape sensitivity, develop the capacity to think critically and creatively to seek solutions and take actions at community and individual levels.

7.14 Critical Thinking and Problem Solving

Applying the perspective of sustainable development raises questions as to what types of knowledge and skills are necessary in order for students to make important choices in life. Schools are institutions within society which play an important role in building values and attitudes. This means that students must be led beyond the study of concepts, principles and theories but additionally to develop view points on the sciences, people, nature and the relationship between people and nature. Sandell et al. (2003: 198) opined that the immediate environment and the local community represent an important starting point for teaching for sustainable development. Environmental problems are transformations that have taken place as a result of human interaction with nature which is in conflict with certain norms and values. In order to build confidence in addressing dilemmas and challenges of sustainable development, textbooks need to address real problems during lessons. These problems need to be relevant to the students' lives and the society in which they live, and preferably problems that the group has some influence over or can affect in some way. Thus an important principle for the choice of content is that depictions of facts should be obtained from numerous sources. This means using alternative perspectives and understandings, including those held by minority groups.

The EVS textbooks attempt to take discussions about the natural environment beyond appreciation of nature to introducing the complexities and issues related to environment and development. These include the social and economic aspects through lessons that focus on displacement due to dams, migration, changes in agricultural practices and their impacts. The textbooks are replete with material which exposes children to such complexities. These are introduced through lesson formats, and techniques which communicate at the appropriate degree of complexity suited to the age level of the students.

Sandell suggests that environmentally ethical reflection is always characterised by the evaluation of alternatives and this type of evaluation requires some system of measurement e.g. a benchmark. In order for students to be able to develop a reflective, environmentally ethical approach in relating to the world, they should be given the opportunity to evaluate critically and discuss the different ways of determining the benchmarks (Sandell et al. 2003: 229). The lessons in the textbooks pose questions that invite such reflection and debates on alternatives, without necessarily suggesting the perfect or even 'right' solution. The textbooks also include a number of stories that talk about positive actions to preserve and protect the environment. Several lessons and exercises seek to build confidence in the ability of children to 'make a difference,' and empower them to question, debate and explore solutions or answers. This calls for a change in the role of the teacher who has no longer simply to transfer textbook information but equally has to clarify, question, pose problems and make suggestions in order to stimulate further discussions. Providing an opportunity where environmental and development issues can be discussed in a democratic arena helps to develop a number of life skills such as critical thinking and systematic articulation of arguments; negotiation, sensitivity and receptivity to a variety of viewpoints and prepares students to become proactive citizens in a democratic society.

7.15 Local Context and Relevance

The Focus Group Paper on Curriculum, Syllabus and Textbooks expressed concern that, in almost all schools of the country, classroom practices were totally dominated by textbooks. The textbook emerged as the single focus of all the knowledge that a child is supposed to acquire at a given stage or class. "Thus 'teaching the textbook' as appears in the source becomes the whole of education. As a result, undue importance is given to the textbook; it has acquired an aura of supremacy and a standard format. It has to be completed from cover to cover in strict sequence; it has developed a language of its own that is difficult to comprehend and is laden with dense concepts" (NCERT 2006a: 37). The EVS textbooks attempt to break from this mould by encouraging much of the 'learning' to take place not between the covers of the textbook but rather by using the textbook in lessons as the starting point to initiate enquiry and investigation by each individual in his or her own context.

As early as 1937, Gandhi pinpointed the advantages of using the real world as the classroom as this was the closest environment in which children are growing. He pointed out that learning inputs are spread over the entire surrounding environment and the society which can easily be harnessed. He also advocated learning-by-doing as the best method of learning. The Delors Report of 1993 also saw education as a social experience "through which children learn about themselves, develop interpersonal skills and acquire basic knowledge and skills. This experience should begin in early childhood in different forms depending on the situations but always with the involvement of families and local communities" (Delors 1996: 23).

Almost 70 years after Gandhi's articulation, the UN DESD International Implementation Strategy reiterated that learning includes what happens in education systems but extends into daily life. Important learning takes place in the home, in social settings, in community institutions, and in the workplace (UNESCO 2005b). The EVS textbooks reflect this wider perspective of learning spaces and communicate the important pedagogical principle that the world around the student and his or her interactions with it is an endless and rich resource for learning. They invite children to talk to parents, elders and community members to find out about their experiences. This might include: What games did they play when they were young? What kind of house did they live in when they were 9–10 years old? How did they travel, and where did they go? The children are encouraged to go to different sites in the neighbourhood to observe, record, interview and investigate; this could be a construction site, a neighbourhood park, a farm, a market place. All these help the teacher to contextualize the learning within the local and the familiar. The opportunity to involve parents and the wider community in the learning process forges the understanding that there are multiple sources of knowledge which need to be valued and respected. The combination of conventional teaching and out of school approaches enable children to experience the three dimensions of education: the ethical and cultural, the scientific and technological, and the economic and social as underlined in the Delors Report (1996: 17).

The lessons invite children to share their personal contexts and narrate their own experiences – questions that link the lesson to the individual help in this. Questions such as: Can you ride a bicycle? Who taught you to ride? Have you ever been in a thick jungle? How did you feel? Do you know anybody who thinks like Akshay's grandmother? What do you think Akshay should do? Such engagement helps each child relate at a personal level to lessons and not just view these as something to be memorised and reproduced for examinations. The opportunities for personal learning experiences, as well as understanding those of others, help children to construct meaning of the world around them through their daily life experiences. These also help to link textbook knowledge and its application in the child's own context, as well as to understand that there may be no single correct answer or understanding. Scott and Gough (2003: 99), citing Pimbert and Pretty, believe that as knowledge and understanding are socially constructed, they are functions of each individual's unique context and past. There is therefore, no single 'correct' understanding. What we take to be true depends on the framework of knowledge and assumptions we bring with us.

7.16 Multi-method: Variety of Teaching-Learning Activities

The NCF 2005 focused on the processes by which learners actively construct knowledge. The books are based on the recognition that children construct knowledge through their own experiences and by direct engagement in activities. This reflects what Scott and Gough describe as the qualitative tradition of educational thinking which sees learning as an iterative, reflexive process through which the individual progressively constructs meanings out of past and present experience (Scott and Gough (2003: 91). In order to construct their own knowledge, learners need to be actively engaged – both mentally and physically. The Focus Paper on Curriculum, Syllabus and Textbooks further articulated that "Learners can be actively engaged only when they are motivated to learn. Active engagement involves enquiry, exploration, questioning, debate, application and reflection leading to theory building and the creation of ideas/positions. This implies that it is important to create opportunities for questioning, enquiry, debate, reflection and arriving at concepts or at creating new ideas" (NCERT 2006a: 33). Based on this premise the EVS textbooks include a wide range of teaching-learning methodologies to involve children in relating to their immediate environment (from family to community and natural to human-made) through a variety of activities.

The thematic organization of the syllabus provides scope for using a wide variety of formats for lessons. These include stories, narratives, interviews, diaries, news reports, poems and discussion, drawing from, and building upon, the child's experiences and perceptions. The different presentation formats of the lessons, including the variety of visual material, support the theory of multiple intelligences which manifest as different learning styles of children. Some children may be more attracted by the visuals; some may make personal emotive links with the narrative; some may enjoy the challenge provided by the exercises (develop cognitive skills), and some may be attracted by the science/history/geography/language aspects. The lessons have something for everyone, thereby supporting a more inclusive learning environment.

The diverse sources of the content (including newspaper reports, films, true life experiences of ordinary people, personal interviews and folklore) and the variety of formats used for presenting the content itself open up great scope for children's curiosity, vocalization, empathy, experimentation, exploration and discovery, and free expression. The exercises are integrated into the lessons rather than placed at the end of the lessons. They pose imaginative questions and provide space for reflection. These are seen as integral to the process of understanding the key ideas in the lesson, and building links with each one's personal context and experience. The suggested activities help develop a range of skills including observation, recording, written and oral expression, classifying and categorising, and psychomotor skills as well as creative and aesthetic sensibilities. The objective of the activities and exercises is not so much to evaluate the students' knowledge but to provide an opportunity for students to express themselves. The activities are designed to serve not only as evaluation tools but also as opportunities for group and peer learning, and sensitization to diversity, as well as similarities, within the social and cultural context.

The changed form and broadened role of the textbooks reflect a significant trend in many countries over the past few decades. This development implies a redefinition of the role of the teacher. As discussed in the *UNESCO Guidebook on Textbook Research and Textbook Revision* (2009), lessons in which the pupils are placed in the spotlight of the instruction procedure are replacing the traditional teacher-centric classroom. The various materials and methods provided by textbooks are used to inspire pupils to ask questions and to explore ways in which these questions may be answered (Pingel 2009: 49).

7.17 Understanding Plural Realities and Celebrating Diversity

Pluralism is the key to ESD. It does not refer to diversity alone but to the energetic engagement with diversity. It is important that students learn not only about their own environments, which are familiar to them (which can be the starting point), but also about how different people live, why there are differences, and what are the similarities despite the differences. India, being a geographically vast and culturally diverse country, is replete with variety and plurality of all kinds – bio-geographical, environmental, regional, socio-economic, cultural, linguistic and more. The EVS textbooks attempted to reflect these multi-dimensional, plural realities of the country.

The themes around, which the EVS textbooks are organised (Family and Friends, Food, Shelter, Water, Travel, the Work We Do), offer ample scope for exposure to, and understanding of this diversity. The themes as well as the formats of the lessons in the textbooks deal with several social and cultural issues. For some children these may be a part of their daily life (such as issues related to access to food and water and family processes), while for others these may not be things that they had ever thought of. The range of lessons expose children to the wide variety of living conditions – from a makeshift shelter on a pavement to affluent lifestyles in big cities. The textbooks deal with such issues and topics as: the challenges in understanding differences in food not just as flavours but in terms of availability and nutrition; family structures and relationships; the kind of work people do; and the 'value' attributed to different kinds of work and professions. Such lessons provide opportunities to highlight issues of gender, caste, or religious stereotypes. The issues are explored through real-life examples and exercises that invite the children's own observations and experiences. The narratives provide 'guided space for the learner to reflect'.

In a multi-lingual country like India, language represents significant aspects of the culture. While talking about lives of people from different regions within the country, the local flavour is maintained by retaining words and phrases in the original. These arouse curiosity and provide children with reference points to understand and relate with relationships, objects, concepts and processes from different socio-economic and cultural settings. Pluralism also allows different perspectives and opinions on the issues to be discussed and studied. It is in the discussions that take place in the classroom where these different perspectives and opinions and their implications are critically reviewed and valued. The textbook lessons and exercises encourage students to express their opinions and impressions in the classroom. Nurturing open and free expression while, at the same time, respecting the views and experiences of others is an integral objective of ESD.

7.18 Reflecting the New Paradigm

The EVS textbook development team took up the challenge of capturing the spirit of the NCF 2005 and giving concrete shape to the concepts delineated in the syllabus in the form of the textbooks. Though seemingly simple, the agenda was complex. The agenda was to be achieved within the context of the change of paradigm of education from one which stressed rote learning and memorization of facts to process-oriented learning which stressed skills and competencies. In his chapter about "checking the general quality of a textbook", Pingel (2009: 78) elaborated, under these premises the "content could no longer stand alone; it had to support also a range of co-scholastic objectives such as developing critical thinking; discussing different value judgements; building up an understanding of time and space; and developing a sense for taking over responsibility in the community".

The NCF 2005 and the EVS textbooks reflect this paradigm shift in education. They permit the child's view to be the centre of teaching. The child as the learner inhabited the minds of the team developing the textbooks. This focus helped the team to achieve many objectives. The EVS textbooks gave great importance to the 'milieu' in acknowledging that knowledge and understanding are a reflection of the context and that there is no single understanding. This was recognition that each learner and his/her context are unique. This approach allowed for the integration of the diversity – socio-cultural, economic as well as environmental – and the presentation of such integration, as an important value.

The format and approach of the textbooks redefined significantly the role of teachers, shifting the paradigm from teacher-centeredness to learner-centeredness. This was a significant leap. By their very nature, the textbooks inspired students to explore their own surroundings, encouraged them to bring their observations and experiences to the classroom, and motivated the exchange and sharing within and outside the class. This helped to ensure that learning just did not happen out of the text in the textbook, thereby negating the dominance of the textbook as stated in the NCERT position paper (2006a: 37).

The ethos of NCF 2005 – child centred pedagogy; global concerns regarding quality of education; the pluralistic demands of a diverse democracy, and the challenges of a globalized world – had to be infused in the process of developing the textbooks as much as in the textbooks themselves. The effort was that this should become a perspective that informed the treatment of chosen themes rather than a forced presence. Texts sought to avoid the usual trap of packing content as knowledge and skills. The focus was on the aspects of the attitudes and values as these are significant determinants in making choices and choosing the path of sustainability. Kumar states, however, that this is easier said than achieved. "Child centeredness can hardly be disseminated as a slogan, nor can joyful learning take place unless teachers are given a theoretical understanding and self confidence to sustain the recommended pedagogy, and not merely exhorted or pressurised to follow it for the sake of certain outcomes" Kumar (2004: 21). These concerns were addressed through a series of teacher orientation workshops across the country following the introduction of the new textbooks. The workshops, however, threw up their own challenges in terms of teacher receptiveness and understanding of the changed paradigm and its practical implication in classroom transactions. The author was a part of this process as well, but discussing that is beyond the scope of this chapter.

The thinking, and chronology of developments in the area of quality education, as well as the recognition of education as a key driver for sustainable development, both at the national as well as international levels, help to support the argument that quality education is indeed ESD. This close overlap is reflected by Scott and Gough (2003) who write that "sustainable development is unlikely ever to have more than a token presence in school, higher or vocational education unless it can show convincingly that it is a means to improve the quality of education generally – not according to the standards of environmentalists, but according to those standards which the main stakeholders in education and learning, such as parents, teachers, pupils, government ministers and employers have arrived at for themselves" (p. 101). India's NCF 2005 is based on this recognition, and the EVS textbooks developed within this framework are an attempt to infuse the sustainability perspective in both theory and practice.

References

Agenda 21, Chapter 36: Promoting education, public awareness and training. http://www.gdrc.org/sustdev/un-desd/c36-a21.html. Accessed 5 Mar 2012.

Daulat Singh Kothari, The architect of defence science in India. Vigyan Prasar Science Portal. http://www.vigyanprasar.gov.in/scientists/dkothari.htm. Accessed 15 Feb 2012.

Delors, J. (1996). *Learning: The treasure within, report of the international commission on education for the twenty first century*. Paris: UNESCO.

Green Teacher. (2011). http://www.greenteacher.org/?page_id=69. Accessed 19 Feb 2012.

Kumar, K. (2004). *Quality of education at the beginning of the 21st century: Lessons from India (background paper). EFA global monitoring report: The quality imperative*. New Delhi: UNESCO.

MHRD (Ministry of Human Resource Development). (1993). Government of India *Learning Without Burden*. New Delhi: MHRD.

MHRD (Ministry of Human Resource Development). (2011). *National policy on education*. http://education.nic.in/NatPol.asp. Accessed 22 Feb 2012.

Naik, J. P. (1975). *Equality, quality and quantity: The elusive triangle in Indian education*. http://www.ibe.unesco.org/fileadmin/user_upload/archive/publications/ThinkersPdf/naike.PDF. Accessed 20 Feb 2012.

National Council of Rural Institutes (NCRI). (2009). *Nai Talim*. http://www.ncri.in/html/naitalim.html. Accessed 5 Mar 2012.

NCERT. (2000). *National curriculum framework for school education*. New Delhi: NCERT.

NCERT. (2005). *The national curriculum framework*. New Delhi: NCERT.

NCERT. (2006a). *Position paper, national focus group on curriculum, syllabus and textbooks*. New Delhi: NCERT.

NCERT. (2006b). *Position paper national focus group on habitat and learning*. New Delhi: NCERT.

NCERT. (2006c). *Syllabus for the classes at the elementary level*. New Delhi: NCERT.

Pingel, F. (2009). *Guidebook on textbook research and textbook revision* Paris: UNESCO: Georg Eckert Institute for International Textbook Research.
Rao, M. V. R. (1999). *Public policy formulation: A study of national policy on education.* Unpublished thesis, Department of Political School of Social Sciences, University of Hyderabad.
Sandell, K., Öhman, J., & Östman, L. (2003). *Education for sustainable development, nature, school and democracy*. Lund: Studentlitteratur.
Scott, W., & Gough, S. (2003). *Sustainable development and learning*. London/New York: Routledge Falmer.
Scoullos, M., & Malotidi, V. (2004). *Handbook on methods used in environmental education and education for sustainable development*. Athens: MID-ECSDE.
UNESCO. (2005a). *Draft international implementation scheme*. Paris: UNESCO.
UNESCO. (2005b). *International implementation scheme* (IIS). Paris: UNESCO: United Nations Decade of Education for Sustainable Development.
WCED (World Commission on Environment and Development). (1987). *From one earth to one world: An overview*. Oxford: Oxford University Press.

Chapter 8
ESD in the Small Island State of Maldives

Mohamed Shareef

8.1 Introduction

The Maldives is an archipelago in the Indian Ocean. 'Sun, sea and sand' form the foundations for its economy. The Maldives is endowed with 1192 beautiful tropical islands and a marine environment, pristine and rich in biological diversity. These unique natural attributes, coupled with investments in human capital and produced capital have made the tourism industry grow steadily since the 1980s when tourism was institutionalized by the government. Tourism is now the mainstay of the Maldivian economy (Ministry of Planning and National Development 2007). It also has a vast Exclusive Economic Zone (EEZ) of 850,000 km^2 in the Indian Ocean which provides rich fishing grounds (Ministry of Planning and National Development 2007: 5). Certainly, the archipelagic character also shows how fragile Maldives is. The Maldives is a very low-lying nation, merely 3–5 ft above sea level – the danger of total submergence due to rising sea levels is quite high.

Ecologically, the Maldivian islands are at the stage of island formation where the crust or the original volcanic core has been completely submerged, leaving only the collection of unconsolidated coral sand over the reef above the surface of the ocean. Consequently, these islands are extremely dynamic systems. Even without anthropogenic interference they are often unstable, as stated by a research study on *Tourism and Environment: Current Issues* (http://www.fao.org/docrep/X5623E/x5623e0p.htm#TopOfPage. Accessed May 1, 2012). The very small size and virtual isolation of the islands makes their ecosystems both on land and sea particularly very fragile. The Indian Ocean tsunami of 2004, the worst natural disaster in the history of the Maldives, brought home the truth that natural disasters are an impending threat along with other vulnerabilities to the small island state. The tragedy

M. Shareef (✉)
Leturer, Faculty of Education, Maldives National University, Malé, Maldives
e-mail: mohamed.shareef@mnu.edu.mv; mohamedshareef@yahoo.com

R.N. Gorana, P.R. Kanaujia (eds.), *Reorienting Educational Efforts for Sustainable Development*, Schooling for Sustainable Development,
DOI 10.1007/978-94-017-7622-6_8

affected the entire country. It shattered livelihoods and caused widespread trauma and distress. It resulted in a fall in GDP of 4.6 % in 2005 (Ministry of Planning and National Development 2007).

Climate change poses a great threat to this small island state. The very physical existence of this nation could be in jeopardy as the islands would very easily be inundated by sea levels in excess of 1 m above current levels – levels that could be reached by 2100 (AOSIS 2012). With climate change becoming the big crisis that the world is trying to deal with, the Maldives has emerged as a vocal member of the Alliance of Small Island States (AOSIS). AOSIS is an intergovernmental organization of low-lying coastal and small island countries. Established in 1990, the main purpose of the alliance is to consolidate the voices of Small Island Developing States (SIDS) to address global warming. "It is well known and confirmed by the IPCC in all of its Assessments that small island developing states (SIDS) – whether located in the tropics or higher latitudes – have characteristics that make them especially vulnerable to the effects of climate change" (http://aosis.info/sids-dock/. Accessed May 1, 2012). These characteristics include their limited size, geographical dislocation, proneness to natural hazards and external shocks, high exposure of population and infrastructure and limited adaptive capacity. The vulnerabilities resulting from these characteristics are exacerbated by the effects of climate change – which include rising seas, acidification of oceans, coral bleaching, coastal erosion, flooding, loss of fresh water supplies, biodiversity loss and more frequent and intense weather events, including hurricanes.

Bhandari and Abe (2001) state that despite being constrained by a lack of technical know-how, manpower and finances, the Maldivian government has instituted policies, legislation and regulatory institutional measures for environmental planning and management, disaster preparedness and regulated tourism. The Government has placed an equally high priority on Environmental Education (EE) and awareness and a number of initiatives have been taken. For this reason the government of Maldives remains strongly committed to advancing EE (Shareef 2010).

8.2 Education, Environment and Development

The Maldives moved to a unified national system of education in 1978. The vision of the Maldives Ministry of Education is:

> Every Maldivian will have access to quality primary and secondary education with widening opportunities for higher education and training. The education provided will promote maximum realization of individual potential while at the same time instil religious and cultural values that will foster social unity and harmony, and, love and pride in the nation (Srivastava 2012: 27).

There has existed a traditional system of schooling for a long time in the Maldives and it was privately owned and operated by island communities. This system has contributed towards achieving many educational objectives, including a high rate of literacy and the preservation of national culture and tradition (Ministry of Education 1992). The government of Maldives spends 8 % of its GDP on education; this figure is high when compared to other countries in South Asia, where, apart from Bhutan, expenditure on education usually ranges from 2 to 4 % of GDP (http://www.infodev.org/en/Publication.880.html. Accessed May 3, 2012). The Maldives has achieved its Millennium Development Goal (MDGs) of universal primary education with a hundred percent enrolment rate at the primary level and is on its way to achieving universal secondary education. Government policy ensures that all islands offer primary education up to Grade 7. School education in the Maldives is structured on a 5-2-3-2 cycle. Primary education begins at age seven (Grade One) and continues till age 13 (Grade Seven). These 7 years of primary education are compulsory for all. Primary education is followed by 3 years of lower secondary studies and 2 years of higher secondary studies. A significant achievement is that the country has been able to maintain a low drop- out rate, resulting in a high gross enrolment rate at the secondary level as well.

The vulnerability of the Maldives islands to environmental change has influenced ESD policy developments in the country. Environmental education (EE) is an important strategy in addressing environmental problems faced by communities around the world (Kwan and Stimpson 2003; Ravindranath 2007; Venkataraman 2008). The concept of environmental education (EE) emerged in the late 1960s. In 1972, during the Stockholm UN Conference on Human Environment, it was recognized as an important tool to promote the protection of the environment and later was acknowledged as the pre-requisite for any development, as recorded by Scoullos (2010: 47) and Plamer (1998) "the case for the promotion and development of EE on a major global scale" is the ever-increasing threat to the resources of the earth and to the health and stability of societies which "justifies an urgent need for an informed global citizenship" (p. 35). This commitment to environment and sustainable development is reflected in important policy documents of the Maldives with a great deal of stress on strengthening the capacity to plan and manage EE and public awareness programmes.

Scott and Gough (2003: 12): remarked that SD and learning are both separately of considerable interest to policy makers with a note that what is really of 'interest' in particular cases may vary a great deal. What is really of 'interest' to the policy makers of the country is defined by the fact that the Maldives is an island nation with its economy dependent on tourism and fisheries. A National Adaptation Plan of Action prepared by the Ministry of Environment, Energy and Water listed the vulnerabilities to and impacts of climate change on the eight main sectors: land, beach and human settlements; critical infrastructure; tourism; fisheries; human health; water resources; agriculture and food security; and biodiversity.

In 1989, the first National Environment Action Plan (NEAP I) was developed to help the government maintain and improve the environment of the country for the collective benefit and enjoyment of the present and future generations. The influence of *Our Common Future* (WCED 1987) can be seen in the inclusion of 'future generations' in the action plan. This was an effort to secure the mainstreaming of sustainability in broad development policy objectives. NEAP II was formulated in

1999. The Third National Environment Action Plan (NEAP3) set out the agenda for environmental protection and management in the Maldives for the 5 year period from 2009 to 2013. Principle 1 of the Rio Declaration on Environment and Development emphatically asserts that "Human beings are at the centre of concerns for sustainable development" (UNCED 1992: http://www.un.org/documents/ga/conf151/aconf15126-1annex1.htm/. Accessed May 3, 2012). The NEAP3 is targeted to achieve measurable environmental results that matter to the people of the Maldives. The six strategic results that are expected are: resilient islands; rich ecosystems; healthy communities; safe water; environmental stewardship; and a carbon neutral nation. It clearly states as one of its principles that "Protection of the natural environment and practising environment friendly lifestyles is a responsibility of every Maldivian" (MHTE 2009a: 4).

In 2009, the government of Maldives adopted a National Sustainable Development Strategy (NSDS) with the help of the United Nations Environment Programme (UNEP). Developmental planning, probably for the first time, integrated ecological perspectives, gave importance to traditions and culture and worked for development in a campaign mode. NSDS provides a long-term vision of sustainable development with clear goals and objectives taking cognizance of national to sub-national aspirations and realities. Probably, the innovative feature of NSDS is that it is aimed to ensure convergence, complementarities, coherence and coordination between existing policies and development frameworks. "Four steps are important for the NSDS in the Asia Pacific context: adopt eco-development principles at the centre of developmental planning; valuation and respect for eco-system services; bring in culture and traditional value at core of development; and develop a campaign for sustainable development" (UNEP: Generic guidelines for developing the National Sustainable Development Strategy 2006: 14).

The NSDS sets out the strategy on how the Maldives will fulfill its commitment to meet the challenges of sustainable development. The overall aim of the NSDS of the Maldives is to identify and develop actions to enable the people of the Maldives achieve continuing improvement in their quality of life both now and in the future. The NSDS established specific goals, objectives and targets that the government, businesses and community must achieve together (MHTE 2009b). While the country moves forward, achieving economic growth and social development, unique environmental challenges confront the nation in its endeavour towards sustainable development.

8.3 How EE/ESD Has Evolved in the Country?

As in many other developing nations, the goals of education are clearly linked to economic development. The government recognizes the "importance of an educated citizenry with the skills and flexibility to respond to employment opportunities" (Bhandari and Abe 2001: 147). So the stress was on technical and vocational education.

In the Maldives, national policies on environment and education recognize 'protection of environment' as a value which must form an integral part of the curriculum. In 1984, based on the recommendations of a national workshop held in 1982, the Ministry of Education introduced environmental studies as a course in the National Curriculum for primary level that is from Grades 1–7. The course was revised, due to deficiencies, in 1990 with the inclusion of wildlife conservation, environment friendly behaviour, climate change and sustainable development (Bhandari and Abe 2001). Environmental education is the main avenue which could be oriented and constructed as an effort for the incorporation of ESD across the formal curriculum. EE is a mandatory subject in the primary school curriculum. On average there are six periods of environmental studies (es) per week at the primary level, which is equal in number to that for mathematics and English.

EE and ESD need not be limited to formal education systems. Formal, non-formal and informal modes offer the means to foster carefully learning that links environment and development not as two sides of a coin but as conjoined spheres. The responsibilities for EE have been delegated by the government to a range of agencies including the Ministry of Education and the Ministry of Environment, Energy and Water (MEEW). Such administrative systems play an essential role in serving as conduits for translating various policies related to economic progress, environmental protection and national identify.

The Educational Development Centre (EDC) of the Ministry for Education is involved in the whole spectrum of educational development work including curriculum development, educational material production, teacher training, non-formal community education, educational broadcasting and school construction and upgrading. The Ministry of Environment, Energy and Water (MEEW) has a unit directly responsible for awareness for topics such as waste management, environment impact assessment, environmental regulations, protected areas and the Maldives 'Greening Programme'. Programmes are carried out through school environment clubs and the production of environment awareness raising materials, such as leaflets, posters, banners and calendars, and through activities organised to mark World Environment Day. The mandate includes planning and implementing public awareness programmes and disseminating information about the environment. The Marine Research Centre in the Maldives was set up in 1984 with a mandate to create awareness among people and the government on fisheries and utilization of other marine resources. This centre has a small educational awareness unit to disseminate its research findings and activities. This is undertaken mainly through its library and museum both of which are popular with student groups.

The promise of non-formal education lies in the fact that it can be an innate way of contextualizing learning across age groups, leading from awareness for action to care for the present and the future. "The Non Formal Education Centre is mandated with the responsibility to conduct courses to increase awareness and teach skills for youth and adults, and produce newspapers, magazines, posters, and other such material to increase awareness (Pandya et al. n.d.: 15). According to a report published by Live and Learn, an International NGO working in the Maldives, the focus of past environmental education approaches in the Maldives has been awareness and mostly knowledge-based (Live & Learn Environment Education 2006: 34). The report further identifies that environmental values such as hygiene are emphasized in the schools. It was observed that large posters and boards with hygiene and environmental messages are placed around the schools. The school environmental clubs operate at the discretion of the school and the teachers in charge. An interesting fact is that occasionally whole island communities become involved in activities organized by school environment clubs as if in answer to the universal question of what is to be sustained and for whom? The whole community involvement for such occasions proves environment is to be sustained for human well-being. At the community level the school environment clubs organise World Environment Day activities, such as tree planting, cleaning of the islands and reefs and organising poster exhibitions. Students are engaged in hands-on, active learning that increases their knowledge and awareness about the environment. Each year the prestigious presidential award known as the 'Green Leaf' award is given to the school which makes an outstanding contribution towards the protection and preservation of the environment and promoting environmentally friendly lifestyles.

The environmental studies syllabus for primary level is divided into five units: the people, the earth, living things, the changing world and interdependence. These units are sub-divided into various topics which include student-centerd activities. The activities are designed in such a way that the students play an active role in collecting information, analysing and presenting it and in some cases even in self-evaluating. (Pandya et al. n.d.) states that most of the topics covered in the environmental studies text books in the Maldives are universal science concepts, and are presented as such. The 'activities' generally require students to fill in tables, or carry out simple experiments. In most of the lessons there are no examples from the Maldives to illustrate concepts. The report acknowledges that teachers are encouraged to link textbook topics with their immediate environment, and to involve students in "learning by doing" with respect to the textbook concepts (p. 18).

In the Maldivian education system, a lot of pressure is imposed on the schools to prepare students for the external examination that they sit at the end of secondary schooling. Attaining good results in the Cambridge GCSE examinations is a highly regarded achievement. It is anticipated that this trend has a direct impact on the teaching methodology used by teachers. Teachers regard their prime duty as being to transmit effectively a body of knowledge to children and that, consequently, a teacher must be a content expert. Teachers assume considerable responsibility for the degree of learning success of their students and teachers tend to teach using the same methods and in the same ways they were taught at school. To help children become more actively engaged in their education, the Ministry of Education and UNICEF have been working to transform the education system in the Maldives from teacher-centred rote learning to child-centred active learning (UNICEF 2006).

8.4 An Inquiry–Based Learning Model for ESD

A review of the literature indicates that a pronounced discrepancy exists in many countries between the problem-solving and action-oriented goals associated with the contemporary philosophy of EE and the education that takes place in schools. A number of authors have observed contradictions between the problem-solving and action-oriented goals associated with the contemporary philosophy of EE and an emphasis on the acquisition of environmental knowledge and awareness in school programs (Blumstein and Saylan 2007; Cotton 2006; Stevenson 2007). According to Stevenson (2007), this rhetoric–reality gap is to be expected given the traditional purpose and structure of schooling: "Educational sociologists have described the contemporary role of schools as still primarily concerned with the transmission of cultural knowledge, skills and values" (p. 145).

Education, Kumar (2004: 2) observed, is something that adults want to give children. Traditionally, educational content is structured as subjects organized in watertight compartments as a set of concepts and facts. Such a knowledge-based arrangement and organization suited the examination driven formal education systems which made teaching a knowledge dispensing activity and learning a mere accrual of concepts and facts. There was no significance attached to learning as a process, the output was important – probably a good score in the examination at the end of the term. Thus a clinical neatness in terms of educational processes like an industrial process has set in and probably succeeded in producing archetypes.

Education for sustainable development requires the reorientation of many existing education policies, programmes and practices to address the social, environmental and economic knowledge, skills, perspectives and values inherent to sustainability (UNESCO 2006). It is now accepted in principle that EE should be incorporated into formal education policies and practice. "EE should be integrated into the whole systems and formal education at all levels to provide the necessary knowledge, understanding values and skills needed by the general public and many occupational groups for their participation in devising solution to environmental questions" (UNESCO-UNEP 1977: 24).

In the Maldives National Curriculum Framework developed by the Educational Development Centre (EDC), environment, science and technology has been identified as a key learning area. It focuses on exploring the natural world and its phenomena, how it relates to the environment and society through systematic and organized inquiry. The Framework has also identified learning for sustainable development as a key competency. This competency aims to raise awareness of the necessity to engage in sustainable practices and to learn the skills of responsible conservation for the future. It encourages young people to treat all life with respect and sensitivity, and manage their resources well. The knowledge, skills and attitudes that they develop through this competency will help them to satisfy their basic needs and have a quality life without jeopardizing the life of future generations. The Framework which is an important policy document clearly signals the commitment of government to support the incorporation of ESD in the formal school curriculum.

EDC is in the process of reviewing primary environmental studies text books to include more environmental concepts and inquiry-based learning activities. Several studies have found that inquiry-based teaching and learning methods positively affect student performance (Allan and Norman 2007; Karaduman and Mehmet 2007; Lord and Terri 2006). Referring to inquiry-based learning, Exline in the Foreword to the book *Integrating Inquiry Across the Curriculum* (Audet and Jordan 2005: v) stated the following with respect to traditional delivery of education: “Unfortunately, our traditional educational system has worked in a way that discourages the natural process of inquiry. Students become less prone to ask questions as they move through the grade levels. In traditional schools, students learn not to ask too many questions, instead to listen and repeat the expected answers”. As an approach to education, inquiry gives full recognition to the relationship between individual and the society. On the one hand it builds upon the experiences and interests of individuals and encourages them to direct their own learning; on the other, it seeks for the socially valued ways of thinking and acting.

EDC has also initiated an EE project under the slogan “Enviro smart-Future smart” aimed at developing instructional materials and teacher training. High quality resource materials are being developed which can be incorporated into classroom teaching. These materials have now been distributed to a hundred primary schools. In the meantime more than 175 teachers from the outer atolls have been trained with the necessary skills required to deliver the new EE curriculum.

The lessons and activities suggested in environmental studies teaching and learning materials in primary schools in the Maldives is based on the inquiry method of learning (Ministry of Education 2009). This methodology of teaching, the inquiry method, is different from traditional classroom-based teaching, where all the answers come from the teacher and the students remain passive recipients of knowledge from a single viewpoint. If we inquire into something, we are directing our thoughts and actions towards developing a better understanding of it. Inquiry-based learning is a process where students are involved in their learning, formulate questions, investigate widely and then build new understandings, meanings and knowledge (Allan and Norman 2007).

According to Murdoch (1998), inquiry as a framework for developing understandings about the world has a long history in educational pedagogy and remains a powerful tool in the contemporary classroom. The inquiry approach reflects the belief that active involvement on the part of students in constructing their knowledge is essential for effective teaching and learning. This means that the role of pupils as constructors of their own knowledge is emphasized. In this process, the teacher acts as a guide, and her or his main task is to support the learning processes of the pupils – such that “teachers and students know together” as quoted by Paulo Freire. “I have a more lucid vision of what we must do to change schooling from the public school we have now, into a school that is happy, into a school that is rigorous, into a school that works democratically. A school in which teachers and students know together and in which the teacher teaches, but while teaching, does not domesticate the student, who, upon learning, will end up also by teaching the teacher”, as spoken in an interview (Torres 1990). As Paulo Freire in another interview with

Aurora stressed, the need for knowledge to be grasped should come from the queries generated by the learners themselves and not the teacher, instead to listen and repeat the expected answers.

Exline asserted that inquiry implies involvement that leads to understanding. Furthermore, involvement in learning implies possessing skills and attitudes that permit the learner to seek resolutions to questions and issues while constructing new knowledge. Useful application of inquiry learning involves several factors: a context for questions, a framework for questions, a focus for questions, and different levels of questions. Well-designed inquiry learning produces knowledge formation that can be widely applied. A complex process is involved when individuals attempt to convert information and data into useful knowledge (Audet and Jordan 2005: v).

8.5 Teacher Education

Conventionally, there has been a strong emphasis placed on examination-oriented, subject-centred teaching in the Maldivian education system (Shareef 2010). Teachers regard it as their prime duty to effectively transmit a body of knowledge to children and that, consequently, a teacher must be a content expert. Therefore teachers assume considerable responsibility for the degree of learning success of their students.

It was declared at the 4th International Conference on EE held at Ahmedabad in 2007 that "learning sites and participation patterns and practices need change to allow for effective environmental education and ESD practices" (http://www.unevoc.net/fileadmin/user_upload/docs/AhmedabadFinalRecommendations.pdf. Accessed 3, May). This requires that existing educational structures should be reviewed and changed, and roles and forms redefined. ESD implies a shift from viewing education as a delivery mechanism, to the recognition that we are all learners as well as teachers as declared at the Ahmedabad Conference.

One of the greatest challenges the Maldives faces in reorienting existing education policies, programmes and practices to address the social, environmental and economic knowledge, skills, perspectives and values inherent in ESD is the lack of trained personnel in relevant fields. At the Ahmedabad Conference it was stated "… emphasise and value the role of teacher education as a catalyst for orienting educators to sustainability practices and real world concerns" (http://www.unevoc.net/fileadmin/user_upload/docs/AhmedabadFinalRecommendations.pdf. Accessed May 3, 2012).

As far as teacher education is concerned, the government's policy in the Maldives focuses on the need for trained local teachers. It is important to note that the majority of teachers in the Maldives have a Diploma of Teaching as their highest qualification. More than half of the primary teachers in the remote islands are untrained teachers (Ministry of Education 2009). As far as EE/ESD is concerned, it appears that most teachers (both trained and untrained) may not have the necessary skills and conceptual understandings to cope with the subject. Palmer observed that "unfortunately, no discernible attention has yet been given to train and motivate teachers who would integrate EE into the school curricula" (1998: 221).

EE should be made an integral part of every teacher training programme. Today the Faculty of Education (FE) of the Maldives National University (MNU) is the largest teacher training institute in the Maldives. MNU is also the main institution for tertiary level education in the country. EE and ESD are being incorporated into the structure of Diploma of Teaching and Bachelor of Education programmes offered at the Faculty. In addition to teacher education courses, a Bachelor of Environmental Management course began in 2012 and this is designed to meet the Maldives' future need for skilled professionals in the area of environmental management and education for the environment. It is important to note that a variety of teaching strategies such as experiential learning, inquiry-based learning and cooperative learning, with appropriate means of assessment are taught in teacher education programmes in the Faculty, aimed at equipping teachers with knowledge, skills, perspectives and values related to ESD.

8.6 Disaster Preparedness and Education

The Indian Ocean Tsunami of 26 December 2004, the worst natural disaster in the history of the Maldives, affected the entire country. This tragedy awakened the country to the need for national level strategies for disaster prevention, recognition and preparedness in line with the national disaster preparedness policies (UNESCO 2007: 4). According to this report, on any given day, about 40 % of the total population is directly engaged in the schools. According to the 2005 official statistics, there were 102,073 students enrolled in 334 schools across the Maldives and 5616 teachers teaching students in those schools. Thus, more than 40 % of the total population 270,101 is directly engaged in the schools on any given school day. This provides an opportunity to prepare the young for disaster preparedness.

UNESCO is leading the efforts through the promotion of the Decade of Education for Sustainable Development. In the context of Education for Sustainable Development, the Maldives National Commission is engaged in developing a policy for "School Based Disaster Preparedness Education". To assist Maldivian schools in preparing for and responding to emergencies, the Ministry of Education, with support from UNDP the Maldives has developed the *School Emergency Preparedness & Response Guide* for use as a template in the preparation of Emergency Operations Plans and Procedures for each school in the Maldives. This guide presents specific procedures to be used in preparing for, and responding to, school emergencies. Starting in the academic year 2012, the Ministry of Education of Maldives has initiated implementation of disaster preparedness weeks in schools. The objective is to reduce the damage and loss in schools caused by a hazard through educating the Maldivian student population on prevention and preparedness methodologies that can be undertaken at school level and also enabling students to practice preparedness activities at school level (Pandya et al. 2009) Some remarkable work has been done by Live and Learn, the international NGO. It works with the Government of the Maldives and the Asian Development Bank in promoting sound environmental management in the aftermath of the tsunami disaster in the Maldives. Live and Learn has been involved in developing a rapid assessment of perceptions into the environmental management in the Maldives and developing toolkits on environmental education and community mobilization program.

8.7 The Tourism Sector and Sustainable Development

Tourism development in the Maldives has been based on sustainable exploitation of resources. Natural capital has to be managed so that a stock of natural resources, no less than that which was inherited from previous generations, is left for the use of future generations. In the Maldives, the tourism industry is synonymous with resort islands. Each resort occupies a separate island and is totally self-contained. This isolation is practical because of the availability of a large number of uninhabited islands that can be developed into tourist resort islands. Currently close to a hundred such islands have been developed as tourist resorts.

Tourism is the country's largest industry and contributes the biggest share to the country's GDP. The tourism industry grew to be the most dominant sector in the economy during a period of only 10 years, marking a new epoch in the economic history of the Maldives. The industry increased its capacity from about 200 beds in 1972 to over 11,300 beds in 1982; and stood at 20,800 beds at the end of 2009 (MOTAC 2010). Tourism continues to be the leading economic activity in the Maldives. While one third of the country's GDP is supplied by tourism, it is also the number one employment generator in the country as well as the foreign currency earner for the country. Government, aware of the potential of the industry, institutionalized the industry in the 1980s to avoid non-sustainable exploitation of natural resources. The same principle applies to the intangible resources that are sold as part of the tourist product: the peace and harmony of the society, local culture, arts and crafts, traditions and livelihood. Hence, strict measures are taken to minimize the negative impacts on society and the environment but the economic goals of employment and income generation are not compromised (http://www.fao.org/docrep/X5623E/x5623e0p.htm#TopOfPage. Accessed May 1, 2012).

Major environmental management issues facing both resorts and inhabited islands alike include climate change and sea level rise, scarcity and pollution of fresh water resources, waste management and biodiversity conservation. For this reason education for sustainable development is perceived as a key tool for improved environmental management at the resorts. It is also important to note that the tourism sector is an important stakeholder in achieving sustainable development goals in the country. Tourist resorts are already engaged in addressing environmental issues and assisting island communities.

8.8 The Way Forward

In March 2009, President Mohamed Nasheed announced that the Maldives aimed to become the first carbon-neutral country by 2020. The National Sustainable Development Strategy takes the carbon neutrality goal as the basis of all future policy targets (MHTE 2009b). Special attention is given to inform the public and enable all the sectors to achieve carbon neutrality. To achieve this aim, ESD in the country has to gain momentum. Simply put, people especially the youth, which makes up more than 40 % of the country's population, need to know that everyone can do their bit. The role of education should be strengthened to enhance the sustainability of the livelihoods of all Maldivians. Greater cooperation is needed between the Ministry of Education, Environment Ministry, the Maldives National University, EDC and NGOs to identify local sustainable development challenges, integrate local knowledge and skills into ESD, and exchange experiences and learn lessons for better practice.

Students, in the Maldives, learn environmental studies only in the primary grades unless they specialize in science subjects or geography at the secondary and higher secondary level. However, few students prefer to choose science or geography over the business and commerce stream which gives them better job opportunities in the growing private business sector. This trend makes more urgent the need to incorporate ESD at the secondary level, perhaps as a trans-disciplinary theme.

Teachers are the main professionals who translate the curriculum in the classroom. Therefore, teachers need to be supported through various planned and systematic continuous professional development programmes, exchange of ideas, seminars and cluster teaching at school level. Teachers need to be continually updated with new methodologies, instruments for doing field investigations and experiments. To be effective professional development should be conducted by experts in pedagogy. Resistance to move away from traditional methods of teaching and discouragement while adhering to new methods comes from the lack of a deeper understanding about the methods and techniques. Schools can collaborate with staff at the Maldives National University who can serve as content area experts. Teachers should have easy access to a comprehensive set of instructional materials and necessary tools to implement ESD across all levels of education. The Ministry of Education needs to allocate sufficient resources to support the practice of learner-centred approaches across all schools in the country (Shareef 2010).

References

Ahmedabad Declaration. (2007, November 26–28). 4th *international conference on environment education* held at CEE, Ahmedabad India. Conference proceedings. http://www.unevoc.net/fileadmin/user_upload/docs/AhmedabadFinalRecommendations.pdf

Allan, G., & Norman, P. (2007). Examining the underlying principles of enquiry-based learning: Two instances of where learning sessions start and end. *International Journal of Learning, 14*(8), 157–165. Retrieved February 20, 2009, from EbscoHost database.

AOSIS. (2012). *Small Island developing states*. http://aosis.info/sids-dock/. Accessed 1 May 2012.
Audet, R., & Jordan, L. (2005). *Integrating inquiry across the curriculum*. Thousand Oaks: Corwin Press.
Bhandari, B., & Osamu, A. (2001). *Environmental education in Asia Pacific region* (pp. 47–62). Kanagawa: IGES.
Blumstein, D. T., & Saylan, C. (2007). The failure of Environmental Education and how we can fix it. *Plos Biology, 5*(5), 120.
Cotton, D. (2006). Implementing curriculum guidance on environmental education: The importance of teachers' beliefs. *Journal of Curriculum Studies, 38*(1), 67–83.
FAO. (1997). *Tourism and the environment: Current issues for management*. Ministry of Tourism: Republic of Maldives. http://www.fao.org/docrep/X5623E/x5623e0p.htm#TopOfPage. Accessed 1 May 2012.
Infodev. (2012). ICT4E in India and South Asia – Maldives country study. http://www.infodev.org/en/Publication.880.html. Accessed 3 May 2012.
Karaduman, H., & Mehmet, G. (2007). The effects of constructivist learning principles based learning materials to students attitudes, success and retention in Social Studies. *Turkish Online Journal of Educational Technology, 6*, 98–112.
Kumar, K. (2004). *What is worth teaching?* New Delhi: Orient Blackswan.
Kwan, F., & Stimpson, P. (2003). Environmental education in Singapore: A curriculum for the environment or in the national interest. *International Research in Geographical and Environmental Education, 2*(12), 123–138.
Live & Learn Environmental Education. (2006). *A rapid assessment of perceptions into environmental management in the Maldives* (Vol. 1). Malé: Environmental Education and Community Mobilisation, Ministry of Environment, Energy and Water with assistance from the Asian Development Bank.
Lord, T., & Terri, O. (2006). Moving from didactic to inquiry-based instruction in a science laboratory. *American Biology Teacher, 68*(6), 342–345. Retrieved February 20, 2009, from EbscoHost database.
Ministry of Education. (2009). *The Maldives national curriculum framework*. Malé: Ministry of Education.
Ministry of Education, Maldives. (1992). *Developments in education, 1990–1992: Maldives country report*. Male': Ministry of Education.
Ministry of Environment and Construction, Maldives. (2006). *State of the environment*. Male': Ministry of Environment and Construction.
Ministry of Housing Transport and Environment MHTE. (2009a). *Maldives national strategy for sustainable development, 2009*. Malé: Ministry of Housing Transport and of Environment.
Ministry of Housing Transport and Environment MHTE. (2009b). *Third national environment action Plan, 2009*. Malé: Ministry of Housing Transport and of Environment.
Ministry of Planning and National Development. (2007). *Seventh national development plan 2006–2010: Creating new opportunities*. Malé: Ministry of Planning and National Development.
MOTAC. (2010). *Tourism yearbook 2010*. Malé: Ministry of Tourism, Arts and Culture.
Murdoch, K. (1998). *Classroom connections: Strategies for integrated learning*. Armadale: Eleaner Curtain.
Pandya, M., Joshi, M., & Jain, G. (2009). *Country report Maldives green curriculum*. Ahmedabad: CEE.
Plamer, J. (1998). *Environmental education in the 21st century*. London: Routledge Falmer.
Ravindranath, M. J. (2007). Environmental education in teacher education in India: Experiences and challenges in the United Nation's decade of education for sustainable development. *Journal of Education for Teaching, 33*(2), 191–206.
Scott, W., & Gough, S. (2003). *Sustainable development and learning*. London/New York: Routeldge Falmer.
Scoullos, M. (2010). Education for sustainable development: The concept and its connection to tolerance and democracy. In A. Nikolopoulou, T. Abraham, & F. Mirbagheri (Eds.), *Education*

for sustainable development: Challenges, strategies, and practices in a globalizing world. New Delhi: SAGE Publications India Pvt Ltd.
Shareef, M. (2010). *Environmental education in the Maldives: The implementation of inquiry-based learning at the primary level.* Unpublished Master of Education thesis, The Uitec Institute of Technology, New Zealand.
Srivastava, G. (2012). *Gender and peace in textbooks and schooling processes: The Maldivian experience.* New Delhi: Concept Publishing Company Pvt. Ltd.
Stevenson, R. B. (2007). Schooling and environmental education: Contradictions in purpose and practice. *Environmental Education Research, 13*(2), 139–153.
Torres, C. (1990). Aurora interview with Paulo FREIRE. http://aurora.icaap.org/index.php/aurora/article/view/45. Accessed 23 Apr 2012.
UNCED. (1992). http://www.un.org/documents/ga/conf151/aconf15126-1annex1.htm/. Accessed 3 May 2012.
UNEP. (2006). *Generic guidelines for developing the national sustainable development strategy.* Bangkok: UNESCO.
UNESCO Bangkok. (2007). *Natural disaster preparedness and education for sustainable development, Maldives country report.* Bangkok: Maldives National Commission for UNESCO, Ministry of Education.
UNESCO-UNEP. (1977). *Intergovernmental conference on environmental education.* Tbilisi: UNESCO with UNEP.
UNICEF. (2006). *Rapid assessment of perceptions into environmental education in Maldivian schools.* Malé: UNICEF.
United Nations Educational, Scientific and Cultural Organization. (2006). *Highlights on progress to date January 2006.* Paris: UNESCO.
Venkataraman, B. (2008). Why environmental education? *Environment, 50*, 8–10.
World Commission on Environment and Development. (1987). *Our common future.* Oxford: Oxford University Press.

Chapter 9
Integration of Environmental Education into the National Education System of Nepal

Badri Dev Pande

9.1 Education System and Environment

Nepal's environmental challenges include those naturally induced and those resulting from human activities. A high rate of population growth and rampant poverty in the country has led to the degradation of forest land and forest resources. Abe and Osamu (2001: 158) state that population pressures and poverty have led to inappropriate land use, forest degradation and natural resource depletion which have had an impact on human health, the environment, ecological pressures and the climate. As a result, Nepal is plagued with a multitude of environmental problems, including deforestation, air and water pollution, solid waste disposal, biodiversity loss and soil loss. Climate change is believed to have triggered snow melting in the Himalayas which in turn has intensified glacial lake outburst floods (GLOF). Research has shown that plant species and aquatic lives are also adversely affected due to climate change. In Nepal human induced causes are more responsible than natural causes for deterioration of the environment. Deforestation of land for fuel wood, agriculture, human settlement, road construction and other development activities is reducing forest cover by over 1 % a year. The rivers and streams throughout the country are increasingly being polluted as is the situation with air pollution in major cities. Disposal of solid wastes, especially non-biodegradable wastes, is a challenge for the country's municipalities, the private sector and NGOs. Increased use of drugs and alcohol by youth and adults has exacerbated social disorders including a rise in the incidents of HIV\AIDS.

B.D. Pande (✉)
Senior Faculty Member, National College for Higher Education,
Baluwatar, Kathmandu, Nepal
e-mail: badri_dpande@hotmail.com; badridevpande@gmail.com

R.N. Gorana, P.R. Kanaujia (eds.), *Reorienting Educational Efforts for Sustainable Development*, Schooling for Sustainable Development,
DOI 10.1007/978-94-017-7622-6_9

This chapter describes the initiatives of the Government of Nepal to integrate environmental education (EE) into the national education system. These initiatives include piloting EE for primary schools and integration of EE into the non-formal education programmes. The chapter gives an account of the Government's effort to reorient EE to education for sustainable development (ESD). It also draws lessons learned from these efforts and challenges encountered and suggest measures for responding to these challenges.

Nepal has made remarkable progress over the past 60 years at all levels of education. In 1951, the literacy rate was only about 2 %. The literacy rate of the country reached 65.9 % in 2011. The year 1951 was a turning point in the history of education in Nepal. It was when the 104 years long autocratic rule of the Rana family ended and the country ushered in the path of democracy. From a few hundred schools in the whole country in 1951, the number of primary and secondary schools exceed 33,000 by 2011. The Government of Nepal endeavoured to provide access to education even in remote rural areas through flexible schooling programmes and material incentives. Literacy campaigns are being launched to encourage out of school youth and adults to become literate. The net enrollment rate in these campaigns was 93.7 % in 2009. By law, primary education is not compulsory in Nepal. The challenge is to bring a large number of school aged children to school and a large number of illiterate populations to adult literacy and non-formal education programs.

The Department of Education of the Ministry of Education (MOE) is responsible for the overall administration of school level education in the country. The District Education Office (DEO) in each of the 75 districts is responsible for overall management and supervision of community schools (government funded) in its district. Regulation of private schools also falls within the purview of DEOs. According to the School Sector Reform Plan, 2009–2015, Nepal's school level education system has been restructured as basic and secondary levels. Basic level education spans Grades 1–8 with the first five Grades known as primary and the next three Grades as lower secondary. The last 2 years of the 4-year secondary education is referred to as higher secondary education.

The Curriculum Development Center of the Ministry of Education is entrusted with the development of curriculum and textbooks, and the approval of reference materials. The Ministry's National Centre for Education and Development has developed a teacher training curriculum based on the school level curriculum and provides training to the teachers. The Non-formal Education Centre of MOE develops curriculum for alternative modes of education, designs and develops adult literacy materials and manages literacy programmes including occasional literacy campaigns. Tribhuvan University is the largest university in the country with over 60 campuses and hundreds of affiliated colleges. In addition, there are several other government supported and private universities offering a range of academic programmes.

9.2 Introduction of Environmental Education

Nepal has a rich cultural heritage with the largest percentage of the population following the Hindu religion. Environment is an important thread within the culture. Plant species are preserved for their use in various religious functions and social events. Traditional educational institutions (Hindu gurukuls where Vedic education was provided and Buddhist monasteries where the preaching of Buddha and Buddhist philosophy were taught) taught their students about the importance of maintaining cultural heritage and the use of natural resources as a part of this. Upon his return from his journey to Great Britain, Jung Badhur Rana, the first Rana Prime Minister of Nepal, established a school in his palace in 1854 to teach English language and other subjects modeled after the British system of education. That school is considered as the beginning of modern education in Nepal. Ever since the introduction of modern education 162 years ago, environmental education was taught in the schools in Nepal through such subjects as Nepali language, biology and social studies. It is only recently that separate courses on environmental education (EE) have been made compulsory in Nepali schools. Education for sustainable development (ESD) began with EE.

Following the first Intergovernmental Conference on Environmental Education at Tbilisi (Georgia) in the former Soviet Union in 1977, participants from Nepal organized the first National Workshop on environmental education in 1978. This was the first ever effort to sensitize the policy makers and academic community about the importance of environmental education.

International agencies like the International Union for Conservation of Nature (IUCN), WWF, and the Swiss Agency for Development and Cooperation played a significant role in mainstreaming EE into the formal and non-formal education system of Nepal. Likewise, national NGOs like the Nepal Forum of Environmental Journalists, Environmental Camps for Conservation Awareness and National Trust for Nature Conservation contributed towards raising public awareness and influencing policy makers about the need for EE. ESD issues in Nepal were seriously discussed starting towards the end of the twentieth century. With support from the Development Assistance Committee (DAC) of the Organization for Economic Cooperation and Development (OECD), a national dialogue on sustainable development was conducted in all five development regions of the country. A summary of these dialogues was presented at the regional workshop in Thailand and an international seminar in Bolivia. Based on these the National Agenda for Sustainable Development was prepared for submission to the World Summit on Sustainable Development in Johannesburg, South Africa in 2002. ESD is yet to find its way into the education system in a systematic manner. A national seminar organized in 2003 with support from the Institute for Global Environmental Strategies (IGES) in Japan resulted in the publication of a book entitled *Education for Sustainable Development in Nepal* (Bhandari and Abe 2003). This book contains 13 articles and nine case studies written by prominent personalities in Nepal. The seminar and subsequent publication gave impetus to the need for ESD in Nepal.

With technical assistance from the IUCN, a National Conservation Strategy (NCS) was prepared for the country in 1987 after extensive consultations with key stakeholders including officials from the various ministries, academicians and subject experts. This Strategy was endorsed by the Government of Nepal in 1988. The NCS recognized the need for conservation of the ever degrading natural environment in Nepal. Among the remedial measures to be taken, it emphasized the need for inclusion of environmental education in all levels and forms of formal and non-formal education programmes (IUCN 1988: 96).

Soon after the government endorsement of NCS, the National Planning Commission (NPC) took the initiative of implementing the Strategy. Under the NPC\IUCN NCS Implementation Project a separate unit, Environmental Education and Communication was established with the purpose of integrating environmental education into the formal and non-formal education programmes as well as to raise public awareness about the need for environmental conservation. One of the first activities of the National Conservation Strategy Implementation Project (NCSIP) was to review the existing formal and non-formal education curricula to ascertain the nature and extent of environmental considerations included therein. The review found that the curricula, textbooks and associated materials reviewed contained some basic elements of environmental awareness, but were inadequate to inculcate students with the attitudes and actions necessary to prevent further deterioration of the environment (IUCN 1991: 6–13). The general public was also found to have a low level of awareness of Nepal's environmental issues and problems.

9.3 Piloting EE for Primary Schools

The National Planning Commission (NPC) in Nepal set up a National Steering Committee under the Chairmanship of its Environment Member to provide guidance and directions to environmental education (EE) initiatives. The Steering Committee included representatives of 13 ministries and departments of the Government of Nepal. It decided that EE should begin at the primary level so that knowledge, attitudes and behaviour of the largest group of students are influenced at the earliest stage of their development. A proposal was developed in 1990 for integrating EE into primary education. It was decided to develop a curriculum along with textbooks and resource materials in cooperation with the agencies of the Ministry of Education for eventual integration into the national system of primary education.

As part of the NCSIP activities, EE curricula were developed for the four primary level subjects namely Nepali language, social studies, health education and science. Ministry of Education subject specialists were mobilized to write draft textbooks and teacher directives. These materials were pilot tested in nine selected schools of three districts. The participating primary school teachers underwent 2 weeks of intensive training to familiarize themselves with the basic environmental issues and concerns and to enable them to use effectively resource materials prepared for instruction. The teachers shared their experiences in mid-term and final evaluation workshops where they also discussed problems and constraints as well as possible ways of resolving them. The revised primary school EE curriculum, textbooks and resource materials were published and used by the MOE in integrating environmental concerns into the new textbooks. They served as resource materials for public and private schools as well as for non-formal education programmes (IUCN 1998a, b: 26–33).

9.4 Influencing Educational Policy Makers, Planners, and Other Stakeholders

The 1990s was an important decade for development and expansion of EE. The country had high level participation of the Prime Minster at the Earth Summit at Rio in Brazil in 1992. With technical assistance from the IUCN and financial support from the Swiss Agency for Development and Cooperation, NCSIP was determined to begin integration of EE into the education system. NCSIP prepared draft National Environmental Education Guidelines in 1991 to sensitize educational policy makers, planners, educators and other stakeholders engaged in formal and non-formal education activities. The same year a national conference on environmental education was organized jointly by the National Planning Commission, Ministry of Education, UNICEF Nepal and IUCN and it attracted a hundred individuals from governmental and non-governmental agencies. The Draft National Environmental Education Guidelines was further refined at this 3 day conference. The conference participants also critically reviewed the curriculum, textbooks and resource materials prepared for primary schools. They deliberated on the role of NGOs in conservation awareness in plenary and in group work. A theme paper on *Introduction to Environmental Education and its Status in Nepal* made the following recommendations (IUCN 1992: 51–52):

1. Environmental education in the education system should be consolidated and integrated into a related national programme.
2. Environmental conservation education should be made an integral part of general education at both school and tertiary levels in social science, humanities, natural science and technical education.
3. The focus of school education should be on local level environmental education and management.
4. For students likely to drop out of school, environmental education should be integrated into functional literacy programmes as well as into vocational and technical training programmes for young people.
5. Environmental education aimed at mid-level technical human resource development should focus on environmental hazards in different sectors and should build capability for national management of the environment.

6. Environmental education at tertiary level should focus on making graduates better planners in environment management with more emphasis on conservation.
7. Environmental conservation education in literacy programmes should be followed by a post-literacy campaign.

The Eighth Five Year Plan (1992–1997) of the Government of Nepal considered the recommendations of the National Conference on Environmental Education. Accordingly, the plan targeted to integrate EE into all levels of formal and non-formal education during the period of the plan. A review committee of this Plan recommended the establishment of a link between education and development. It suggested that Nepal should set the goals of:

> Reorienting education for the alleviation of poverty, eradication of illiteracy, improving the access of disadvantaged groups to education, developing human resources in accordance with national development requirements, and making education contextual and relevant to life situations (Bhandari and Abe 2001: 159).

The National Education Commission which was formed by the Government of Nepal in 1991 thoroughly reviewed the education system in Nepal and made recommendations for consolidation and strengthening of educational programmes. In its report submitted in 1992, the Commission recommended for inclusion of EE courses in the school level educational programmes. Soon after submission of the report, the Curriculum Development Centre started to develop a curriculum and textbooks for EE. Thus the government initiatives were all positive towards raising conservation awareness and imparting knowledge and skills for behaviour change through formal modes of education.

In the 1990s, significant development took place in the integration of environmental education into non-formal education programmes. Through the initiatives of the National Conservation Strategy Implementation Project, resource packages were prepared for the six training centres established to train all levels of government employees to enhance their capacity to discharge their assigned responsibilities. Training centres involved were the Nepal Administrative Staff College, the Women's Development Training Centre, the Hotel Management and Tourism Training Centre, the Agricultural Training Section of the Ministry of Forests and Soil Conservation and the Agricultural Training and Research Institute of Agriculture Development Bank of Nepal. With the help of their resource packages, the trainers incorporated environmental concerns into their programmes which were of various durations ranging from 1 week to 3 months. The resource package was divided into four parts the first of which consisted of the basic concepts, issues, problems and ways of resolving those problems. The second consisted of activities both in classrooms and out of classrooms designed to enhance conceptual clarity. The third part of the resource package was aimed at assisting the trainers in evaluating and monitoring the knowledge and practice of conservation education of their trainees. The fourth part consisted of a guide to resource materials including audio-visual materials, publications and models prepared by the various agencies.

A number of factors were responsible for Nepal's success in mainstreaming EE into formal and non-formal education programmes in the 1990s. First, the initiative to prepare Nepal's National Conservation Strategy in 1985 gave a shape to the NCS document in 1987 and the Government of Nepal adopted it as its conservation strategy in 1988. With their technical support, international non-governmental organizations like the IUCN were able to convince the government that a national policy on environment was necessary. The NCS document included a section on the need for integrating EE into formal education and public awareness programmes. Secondly, donor agencies were ready to provide financial assistance for EE initiatives. For example in the preparation of the NCS document, the Canadian International Development Agency, the Swiss Development Cooperation and the United States Agency for International Development extended their hands. SDC continued its support for the preparation of EE curricula and textbooks. Third, high level participation from Nepal in the Earth Summit helped to translate its Agenda 21 into action. The Eighth Five-Year Plan (1992–1997) and the Report of National Education Commission in 1992 recommended integration of EE into the education system in Nepal. Finally, the Ministry of Education and agencies under it and the various training centres of different ministries played a significant role in providing entry to EE in the national system of formal and non-formal education.

9.5 Reorientation of Environmental Education to Education for Sustainable Development

Education for sustainable development (ESD) is a relatively new concept in Nepal. At the initiative of the National Panning Commission and with support from Development Assistance Committee (DAC) of the OECD (Organisation for Economic Co-operation and Development), national level dialogues on sustainable development were organized in 2000 and 2001 on what constitutes sustainable development. Five such dialogues were organized in the five development regions of the country. In each more than one hundred participants from all parts of society expressed opinions on what they felt sustainable development had taken place at the national level and in their own development regions. These dialogues provided valuable inputs to the preparation of a country report for presentation at the World Summit on Sustainable Development in Johannesburg, South Africa in 2002. The report entitled *Sustainable Agenda for Nepal* was published in 2003 by National Planning Commission. The policy directions in the Agenda were grouped under seven areas: economic growth and poverty reduction; population and settlement; forest, ecosystems and infrastructure; and peace and security; for education, the Agenda emphasized vocational training opportunities, improvement in the quality of colleges and universities and improvement in domestic research capacity (NPC 2003: 25–29).

Nepal's School Sector Reform Plan 2009–2015 was supposed to start implementation in its entirety in July 2009. However, pending legislation to this effect, the Plan has not been fully implemented. Once the legislation is through the parliament, the existing curriculum will be revised. The Ministry of Education has already brought out a National Curriculum Framework. New courses of study will be developed following the directions provided by this curriculum framework. Some of these courses are likely to include Education for sustainable development matters as there is already national policy in this regard. EE components are incorporated into both formal and non-formal education programmes currently in place. The Ministry of Education prescribes a uniform curriculum at the school level. Multilateral as well as bilateral agencies and international and national NGOs are extending their support to EE.

9.6 EE in the Formal Education Sector

School level education in Nepal consists of basic education, Grades 1–8, and secondary education, Grades 9–12. As early as Grade One, children are provided with the opportunity to familiarize themselves with the environment in their surroundings. Youngsters are made aware of living beings including animals and plants and the need to show kindness towards animals and to take care of plants. Through their courses of study in social studies and creative art and science, health and physical education, children from Grades 1–3 learn about social values and norms, and cultural heritage. They are made aware of social problems and ways of resolving them. In Grades 4–5, under the subject science and environment, children learn about living beings, the meaning of environment, matter and energy, earth and space, information technology and local technology. A special feature of the primary level curriculum is that one subject can be designed locally either to learn about the local language or on topics relevant to the particular community. This is a good opportunity to gain more insight and practical experience in environmental education and education for sustainable development.

In Grades 6, 7 and 8 a separate course, Environment and Science is offered. From this course the students learn about the impact of environmental degradation, how a balance in the environment can be maintained and what it is meant by environment and sustainable development. The 4 years of secondary schooling is divided into 2-year programmes. Grades 9 and 10 students are offered 100 marks worth of compulsory subject entitled Health, Population and Environmental Education which is the weight assigned to a full subject of study.

There are nine topical units for Grade 9 and seven units for Grade 10. The broad content areas are: health, population and environmental education; family life education; causes of population change, its impact and measures for population management; natural resources; caring for the earth; environmental sanitation; adolescence, sex and reproductive health education; consumer health; and primary health care and safety education. Students are evaluated in both their theoretical (75 %) and practical (25 %) achievements. Secondary level students can enroll in the optional course of Environmental Science during the first 2 years of secondary level education. Besides providing more in-depth and additional knowledge, this course consolidates learning in the Health, Population and Environmental Education course. Environmental education is not compulsory for higher, or the second half of, secondary education. Students may enroll in optional subjects of environmental education in both Grades 11 and 12. Topics covered include: ecology and the ecosystem, environmental pollution, natural hazards, environmental degradation and mitigation measures (IUCN-NCS 1991: p. 7–13).

At the university level various post-secondary education programmes offer courses in environmental education. The 3-year bachelor's programme of the Faculty of Education of Tribhuvan University, for example, offers two optional courses on environmental education to pre-service secondary school teachers. In the Faculty of Science and Technology of the same university, a Masters' programme in environmental science is offered. The Bachelor of Development Studies programme of Kathmandu University provides a compulsory course entitled Education for Sustainable Development. Other universities and colleges in Nepal offer EE\ESD as separate courses or as components of courses in their science and humanities faculties.

9.7 EE in the Non-formal Education Sector

According to Nepal Labour Force Survey 2008, the literacy rate for 6+ years was 60.5 %. Efforts are made to impart literacy and mainstream out-of-school children and youth into the formal education system through alternative courses of study. Employees in the governmental and non-governmental sectors are receiving training in EE especially to make them aware of how their jobs are linked with the environment and what they can do to conserve it. The Non-formal Education Centre (NFEC) of the Ministry of Education has been playing a leading role as the apex body in policy formulation, curriculum and resource materials development and implementation of non-formal education programmes throughout the country. With a focus on literacy, NFEC has launched several programmes including the National Literacy Campaign, adult literacy, women's literacy, school outreach and flexible schooling programmes targeted at out-of-school children. The National Literacy Campaign, although of only 3 months in duration, includes such content areas as environmental protection, afforestation, civic consciousness, and health and sanitation and the safe use of water (Non-Formal Education Centre 2010: 6–7). These content areas are also included in the adult literacy programmes of 6 months duration. NFEC maintains community learning centres throughout the country where community development activities and life skills programmes are organized.

The six government training centres continue to provide conservation education to pre-service and in-service government employees in various areas of development activities. The private sector has also become increasingly aware of the need for taking measures to mitigate adverse effect on the environment resulting from their industrial and commercial operations. Numerous NGOs and CBOs are engaged in raising awareness about the adverse consequence of environmental degradation and encouraging communities and concerned individuals to take action towards sustainable development activities. There are several NGOs of national standing able to make an impact on communities at large. To take a few examples, the Nepal Forum of Environmental Journalists organizes nationwide programmes through radio, documentary films, wall newspapers, and interaction programmes on current and pressing sustainability issues. The National Trust for Nature Conservation maintains offices in different parts of the country where research, training and awareness raising activities are organized on the importance of conservation of natural resources located within national parks and protected areas and in biodiversity rich communities. Environmental Camps for Conservation Awareness (ECCA) is another national level NGO that is particularly known for its conservation education programme. ECCA has trained hundreds of young boys and girls as counselors who run 5-day ECCA camps for school children in their own communities. The children are made aware of their community environment and are taught practical skills such as the construction of fuel efficient stoves.

A number of international non-governmental organizations (INGOs) have played a pivotal role in the promotion of EE in Nepal. IUCN has provided technical assistance to major governmental agencies to formulate environmental education policies and to develop EE curricula and resource materials for formal as well as non-formal education programmes along with their pilot testing. The Worldwide Fund for Nature (WWF) has been supporting conservation education programmes in Nepal for a number of years. School children have been exposed to practical experience through eco clubs. Non-formal education programmes for local women and exchange visits for students and community members have been organized. Trainings, workshops and conservation programmes have been organized for students, teachers, community members and governmental and national park staff.

Multilateral agencies like the World Bank and Asian Development Bank have long been contributing to sustainable development education in Nepal through loans and grants. The World Bank loan fund has been effective in transferring school management responsibilities from the government to communities. These multilateral agencies have supported Nepal's Millennium Development Goals and Education for All programmes. The roles of bilateral agencies such as those of FINNIDA, DANIDA, and NORAD have been significant in supporting these programmes and have agreed to pool their contributions to avoid duplications.

The consolidation of EE and its reorientation towards ESD has been possible due largely to the work and support of these national and international agencies, the bureaucracy in the government being too bogged down with routine work. It is through INGOs and international development agencies that the politicians, government ministries and departments are made aware and motivated to make timely reform in education.

9.8 Lessons Learnt

The lessons learnt from about two decades of ESD experience in Nepal can be summarized as follows:

- Total commitment and involvement of the implementing agencies is necessary to ensure the success of any EE programme. It was the strong commitment and involvement of the Ministry of Education as well as the Curriculum Development Centre that made it possible to introduce new school level EE subjects which all students were required to take. Likewise, it was due to the cooperation of training centre authorities and trainers that made it possible to incorporate EE components into a wide range of training programmes designed for government employees.
- For non-formal education programmes, EE components integrated within an ongoing programme are found to be the most effective. The NCS Implementation Project was able to collaborate with the six government training centres in integrating EE components needing only a small amount of financial investment. The Non-formal Education Centre of the Ministry of Education has successfully integrated life-related EE components into its literacy and post-literacy programmes targeting adult female and male students.
- Many NGOs in Nepal have developed their capacity to complement national level EE programmes with their technical staff and communication infrastructure. They are able to train teachers on practical aspects of EE, disseminate timely information and messages to the rural and urban mass through their FM stations and prepare resource materials to supplement textbooks. Through partnership with governmental agencies, NGOs have been able to consolidate their own experience in the process of providing required services.
- Posters, wall newspapers and street theatres have been found to be the most effective materials and methods for disseminating EE messages. Posters are of great attraction to children and adults alike. Besides disseminating important messages, wall newspapers also serve as post-literacy materials for both the literate and neo-literate. Since messages are combined with entertainment, street theatres are powerful means of communication.
- For effective EE instruction both in formal and non-formal education programmes, it is essential for the instructors to be trained and motivated. In many schools, EE is taught routinely in the same way that history, social science or language courses are taught without meaningful discussions of current environmental issues and problems or any exposure to practical experience.

9.9 Challenges

Although EE is widely recognized as an important subject of study within formal as well as non-formal education programmes, there are a number of challenges to overcome before the expected outcomes of such a subject are realized. Some of these challenges are:

- Important socio-political developments such as the abolition of the monarchy, a shift from traditional farming to commercial farming and increased movement of people within and outside the country have taken place over the past decade. There has been mounting pressure on the Curriculum Development Centre to introduce new subjects at school level. Questions are raised whether separate EE subjects should continue in school level education. The challenge is to retain and consolidate EE subjects while at the same time orienting EE to ESD.
- Unlike in mathematics, science or language related subjects, there are no established EE teachers. There are no specific training programmes for teachers of this subject. As such, it is construed that any teacher who is not overloaded can handle the subject. However, teachers, particularly at secondary level, find it difficult to teach health, population and environmental education combined in a single subject. Although the National Centre for Education and Development has developed training packages and trains teachers to enhance their capacity, more effort is needed to produce EE/ESD teachers.
- There is a general lack of practical experience for students although the curriculum requires this. Students in Grades 9 and 10 are evaluated for their practical experience with a weighting of 25 %. In practice, students rarely engage in practical activities but nonetheless secure good marks because the schools allocate them good marks anyway.
- The same course is taught throughout the country but parts of the curricula are not always relevant to students of a particular geographical location. For lack of practical exercise, many students are introduced to concepts quite irrelevant to their daily life. There is also a need to update the course with a focus on such much discussed topics as climate change.

9.10 Way Forward

- Nepal's rural population is heavily dependent on natural resources for its subsistence living. Over 75 % of the energy needs are provided by fuel wood. The reason behind the popularity of community forestry is that communities are able to derive fuel wood, fodder for animals as well as roofing and construction materials from what was once national forest. Many people depend on non-timber forest products such as herbs, fruits and leaves for making their living. Nepal is considered as a biodiversity rich country with abundant opportunities never exploited before. Through EE\ESD there is a good possibility to benefit more from these opportunities.

- Nepal has undergone major political changes during the past decade. The country has opted for a federal system which is a drastic departure from the monarchy which lasted for almost 240 years. Once the political structure is decided, it is most likely that provinces or the state will focus on the natural resources that are available in each territory. Sensitization of the members of parliament and other political leaders is essential so that appropriate policy measures and legal frameworks are in place to ensure sustainable use of the natural resources of a province.
- ESD can yield better results if it is taught by trained and experienced teachers. A cadre of EE\ESD teachers should be developed as master trainers who in turn should train pre-service and in-service teachers. The role of the National Centre for Educational Development is important in this regard. The practice of bringing in experts from various disciplines and practitioners from classrooms to act as resource persons should be formalized so that students are able to learn from firsthand experience.
- Practical experience should be an integral part of any EE\ESD instruction. School administration should be cooperative and play a facilitators' role in enabling EE\ESD teachers to take their students to demonstration sites and for study tours. Through the formation of eco-clubs and other activities, students should be encouraged to engage in a range of conservation activities.
- Use of resource materials to supplement textbooks should not be limited to printed materials. The surroundings of any school and resource materials available at local level should be utilized as teaching and learning aids to the maximum extent possible.

References

Bhandari, B., & Osamu, A. (2001). *Environmental education in Asia Pacific Region* (pp. 47–158). Kanagawa: Institute for Global Environmental Strategies (IGES).

Bhandari, B., & Osamu, A. (2003). *Education for sustainable development in Nepal – Views and vision*. Kanagawa: Institute for Global Environmental Strategies (IGES).

IUCN. (1991). *Environmental education in Nepal – A review*. Kathmandu: NPC/IUCN NCS Implementation Project.

IUCN. (1992). National Conservation Strategy (NCS) (1992). Implementation project. In *Proceedings of the national conference on environmental education*, Kathmandu.

IUCN. (1998a). *Report on the end of the project workshop on the national conservation strategy implementation project*. Kathmandu: NPC/IUCN NCS Implementation Project.

IUCN. (1998b). National Conservation Strategy for Nepal, NCS (1988). *Building on success*, Government of Nepal, IUCN.

National Planning Commission (NPC). (2003). *Sustainable development agenda for Nepal*. Kathmandu: National Planning Commission.

Non-formal Education Centre. (2010). *Non-formal education centre – An introduction*. Kathmandu: NFC.

NPC/IUCN NCS Implementation Project. (1991). *Draft national environmental education guidelines*. Kathmandu: NPC/IUCN NCS Implementation Project.

Chapter 10
Education for Sustainable Development: Challenges in Pakistan

Tahir Ahmad Dhindsa

10.1 Introduction

The Indian Sub-continent was partitioned in August 1947 to create two independent, sovereign dominiums, India and Pakistan. The idea of partition was based on the theory that Indian sub-continent was made up of Hindus and Muslims. Pakistan officially came to be known as the 'Islamic Republic of Pakistan.' On the eve of independence from British rule, much of the infrastructure in the field of education was left in India. The areas that became Pakistan were relevantly backward and underdeveloped in many respects. About 85 % of the population was illiterate at the time of partition. The rate of literacy among the women in rural Baluchistan was virtually zero. It was necessary to start from scratch. The quality and quantum of all resources be they human or physical were so impoverished that it was deemed impossible for the policy planners to depend on the available human resources for rapid progress on economic and social fronts, let alone employ education as a tool to achieve, what we now understand as 'sustainable development.'

Soon after its independence, Pakistan went to war with India over the disputed territory of Kashmir. From that time Pakistan became a security state, obliged to pay a large standing army and funding of security programmes acquired priority. As a result, development initiatives suffered, and the social sector perhaps suffered more. Nonetheless, as given in the introduction of Vision 2030 adopted by the Government of Pakistan through the Planning Commission, protection of the environment is basic to all development, without which no development is sustainable (Government of Pakistan 2007). The document discusses key targets under the general umbrella

T.A. Dhindsa (✉)
Head of Policy Advocacy & Outreach and Head of SDTV,
Sustainable Development Policy Institute, Islamabad, Pakistan
e-mail: tahirdhindsa@yahoo.com; tahirdhindsa@gmail.com; tahirdhindsa@hotmail.com

R.N. Gorana, P.R. Kanaujia (eds.), *Reorienting Educational Efforts for Sustainable Development*, Schooling for Sustainable Development,
DOI 10.1007/978-94-017-7622-6_10

of the concept of sustainable development, which rests on three intertwined pillars of economic and social development without damaging the environment, as defined in Pakistan Agenda 21.

10.2 Protecting Environment for Sustainability

Although, the subject of environment is an essential component of sustainable development, it has been afforded little academic space and recognition in Pakistan's education curricula. The National Climate Change Policy has, however, put emphasis on understanding and adjusting to climate change (climate adaptation) (Ministry of Environment 2011). http://www.lead.org.pk/cc/attachments/Resource_Center/NAP/pakistan.pdf. Accessed on November 21, 2014. It recognizes that Pakistan is one of the lowest contributors to the global stock of greenhouse gases, but, at the same time, it is also one of the worst affected countries. The policy document recognizes awareness-raising as one of the major initiatives that is necessary. For instance, public awareness-raising campaigns can underscore the importance of conservation and sustainable use of water. As a policy measure, the need for synthesizing knowledge about environment and conservation has been highlighted. However, this has not been translated into curricula. The document may rightly claim that Pakistan is hardly prepared to meet the twenty-first century's biggest challenge. The challenge of climate change needs to be redressed through parallel challenges in the education sector, the most critical problems being limited investment, inefficient training, and 'brain drain'. As a developing nation, Pakistan faces the challenge of educating its population, and strengthening its economy while at the same time bringing due focus to conservation of natural resources and environment using awareness and education. Academia and the policy makers in Pakistan are aware of the need to integrate environmental education in basic curricula but this has not been implemented substantially.

The National Environment Policy 2005 has clearly aimed at integration of environment into development this is need of the hour. "To protect, conserve and restore Pakistan's environment in order to improve the life of the citizen through sustainable development" (Ministry of Environment 2005: 9) is the main goal of the policy. The document in explaining the policy sets out five key objectives in which education and awareness is given recognition. The policy document provides a framework for addressing following environmental issues: Pollution of fresh water bodies and coastal water, air pollution, lack of proper waste management, deforestation, loss of biodiversity, desertification, natural disasters and climate change.

All individual federal authorities were asked to adopt a plan for achieving the goals in the policy document, namely "to protect, conserve and restore Pakistan's environment in order to improve the quality of life of the citizens through sustainable development" (p. 9). To achieve the desired goals the policy objectives were defined; "(1) effective management of environmental resources (2) Integration of environmental consideration in policy planning for making it sustainable (3) Capacity building of all stakeholders for better environment management (4) Meeting International obligations in line with the national aspirations and, (5) Creation of demand for environment through mass awareness and community mobilization" (p. 10).

The policy proposed that the government entered into multilateral agreements with international stakeholders on the environment to pursue overall development goals. It proposed effective participation at regional and international level to establish cooperation for the protection of environment. The policy made it incumbent upon the government to develop and implement national plan in the line with the Johannesburg Plan.

Pakistan feels that it was necessary to reduce poverty for protecting environment and the National Environment Policy 2005, tried to address the issue.

1. "It proposed to integrate poverty-environment issue to the national policy planning.
2. It proposed that monetary allocations should be made to all initiatives taken as such at the grass root level.
3. Capacity at the level of union council should be enhanced to promote community-level environmental management.
4. Bring environmentally friendly appliances like cooking stoves and better crop production technologies to the poor. And
5. Regularise, Kachi Abadies (squatter settlements), and implement National Resettlement Policy, after it is devised" (p. 15).

The policy also provided recommendations on other areas like; gender, health, trade and natural disasters. Instructions were given to make the recommendations functional at the local government level. The policy identified six instruments for achieving the policy objectives. "(1) Integration of environment into development planning (2) Legislation and legal framework (3) Capacity development (4) Economic and market based instruments (5) Public awareness and education (6) Public-private-civil society partnership" (p. 18).

10.3 Education in Pakistan

Since the concept of protecting the environment is new in Pakistan, there is little mention of it in efforts designed to promote education as a fundamental tool for social and economic development. Since independence, about 18 policy frameworks with different titles have been adopted. In between these policy level frameworks, some supplementary programmes like Nai Roshni Schools: 1986, Drop-in Schools: 1986–1990, Nationwide Literacy Programme: 1986–1990, were also introduced. Aim of these crash programmes was to supplement the ongoing policy level programme in the critical areas for better results.

The first elected federal government was formed by Pakistan Peoples' Party in 2009 and introduced many constitutional reforms. This included the 18th constitutional amendment in the constitution which devolved Education to the provinces.

To couple devolution a new National Education Policy was introduced in 2009. Some of the needs identified in the policy document are: (1) Talking advantage of the Demographic Transition, (2) Bringing Uniformity and Confidence in Public Education System, (3) Tackling Challenges of Globalization through Competitiveness, (4) Ensuring Social Cohesion by Inclusiveness, (5) Dovetailing Government Initiatives and, (6) Leveraging International Development Partners. However, little was said for protecting the environment as implementation corollary (Ministry of Education 2009: 9–13).

Based on the state of Pakistan's education, the policy defined five core areas of concern as being central to all deliberations. These five core areas were: (pp. 69–70)

(A) "Access to Education Opportunities
(B) Equity in Education

    1. The Gender Dimension
    2. The rural-Urban Divide
    3. Provincial and Area Disparities

(C) Quality of Provision
(D) The Resource Commitment
(E) Structure of Education: Public-Private Provision"

The policy noted that, "performance remained deficient in several key aspects including access, quality and equity of educational opportunities and the international challenges like Millennium Development Goals (MDGs)..." (p. 7). To provide quality education for all, the policy defined key features, namely:

1. What type of Education
2. Reaping the Demographic Dividend
3. Bringing Uniformity and Confidence in Public Education
4. Meeting the Challenges of Globalization and Competitiveness
5. Converting Social Exclusion into Cohesion
6. Dovetailing Government Initiatives
7. Leveraging International Development Partners
8. Removing Major Deficiencies: Poor Access and Poor Quality
9. Bridging Systemic Gaps:

    (a) The Commitment Gap
    (b) The Implementation Gap

10. Bringing a Paradigmatic Shift.

Many of the targets set in the policy are not greatly different from the Vision 2030. Growth through the knowledge economy and promoting competitiveness through education, are central.

Devolution has led to curriculum development as another major challenge for policy makers. The matter has been totally devolved to the provinces but there are no given standards to complete the exercise. There is no standardized national curriculum, which can be adopted through consensus. On the other hand, policy makers need to recognize social, ethnic and religious difference among the population and find a solution. Pakistan needs to do two things. Firstly there is a need to develop a part of the curriculum which may be adopted as the common fiber throughout the country. This part of curricula should contain basic information and an introduction to Pakistan's geography, recent history and common parts of religion in a comparative way. This should be done under a board that represents all provinces and works under the council of common interests at the federal level. Secondly, at the local level provinces should form their own curricula. However, before graduation to the university level, quality assurance is needed and, this has to take place at the federal level.

Since independence, Pakistan as a country has announced many education policies, but education, as noted earlier, has never been given priority as a policy item. There are a few basic issues. In terms of allocation of resources, Pakistan devotes about 2.1 % of GDP to education, when similar countries like Bangladesh allocates 3 %, India approaching 4 and Malaysia about 8 %. Among its peers, Pakistan spends less. At the same time, governance issues have resulted in poor implementation and compliance with policies. Particularly in the rural areas, the country has had ghost schools with few facilities for students and absentee teachers. Today the net enrollment rate at the primary level is about 60 % and about twenty million students are out of school. This is a story of both low priority as well as poor delivery of education.

Attempts are being made to address the balance of expenditure in the country. Historically, Pakistan has been spending two and half times as much on defense as on education. Just before the announcement of the 7th National Finance Commission (NFC) Award in 2007, spending on education plummeted below 1.8 % of GDP. Today, about 90 % of the public expenditure on education is done at the provincial level and Pakistan is spending 2.1 % of GDP on education, as compared to the previous 1.8 % of GDP prior to the NFC award. However, it still has not improved delivery of service (Provincial Budgets, 2013–2014) and uptake of education. Experts claim that poverty is one of the major reasons behind the stagnation in enrolment rates as increasingly children today are being taken out of schools for child labor. Additionally, terrorism has affected the situation on ground badly. This is especially for girls in the northern part of the Khyber Pakhtun Khwa (KPK) province, where girls' schools have been destroyed and the sending of girls to school declared irreligious by the militants. The damage done to the schools infrastructure is considerable.

Alongside the public sector, the private sector has played an important part in educating the nation. The emergence of a private sector in education is one of the success stories which is recognized outside the country. In the larger cities of Pakistan, private education at all levels is beginning to become as large as public education. The enrollment rate in these cities in private schools is close to 45 %. A range of private schools has emerged, catering to children across different family income levels. Demand for education even in the lower middle classes is such that private schools, even for families existing marginally above the poverty line are viable and teachers are available at all levels. There is a premium now for good education and the private sector has responded to the demand, whereas the public sector has not.

With the passage of 18th constitutional amendment, education has been devolved to the provincial level and side by side, the 7th NFC has provided new resources to the provinces to the extent that these have doubled in the last 3 years since 2010. So there is no dearth of resources; what is really bothering the policy makers is finding new model of an improved service delivery.

10.4 Environment Education in Pakistan

Similar to other South Asian countries, in Pakistan environmental themes are mostly infused in Science and Social Studies subjects at all levels along with a few lessons in the languages. As per the Situation Analysis Study on EE in Pakistan (Shahid et al. 2005) the quality and quantity of content needs revision. The study notes that integration of more activities and improvement in the knowledge content along with the introduction of case studies and environmental issues within textbooks is needed. It envisages that such measures would help in strengthening the National Curriculum for delivering the concepts of EE.

The development of environmental education in Pakistan is piecemeal. It has developed an Environmental Strategy and Education Sub-strategy for the Balochistan and Sindh provinces and includes a key strategy for mass environmental awareness and education. Elsewhere, several NGOs, universities, education departments are involved in planning and implementing EE initiatives in the country which are linked to formal and non formal approach (see inset).

Environmental Education Initiatives Non Government Organizations and Other Institutions in Pakistan

WWF –Pakistan

This is the largest conservation organization in Pakistan and has been working in the field of EE since 1985. It has developed diverse material on EE and has trained teachers throughout Pakistan. A rich source of information on case studies and environmental issues of Pakistan is provided by WWF. Some of the activities undertaken under the projects include: raising public awareness, teacher's training the provision of educational aids, training workshops for communities in conservation, the provision of supplementary materials for curriculum development (IUCN 2003: 14). Apart from this, a Green School initiative is being taken up which focuses on use of innovative teaching methods to sensitize students to discourage pollution, promote water conservation, plant trees and create awareness amongst other fellow students to become eco-friendly (http://wwf.org.pk/blog/2014/10/31/spreading-environmental-education-through-green-school-programme. Accessed on November 21, 2014).

(continued)

IED-Agha Khan University, Karachi

The Institute for Educational Development (IED) offers EE as an elective course at M.Ed level developed with support from IUCN. IED has also integrated EE related concerns in its pre-service and in-service teacher training programmes. They have also worked on lesson plans for EE at primary level. The distance learning course offered by them for educators was developed with support of WWF Pakistan. (IUCN 2003: 18)

Directorate of Education, Northern Areas Gilgit

The Department has worked closely with IUCN and WWF to infuse EE at the school level. They have developed modules and trained a large number of teachers at primary and secondary levels (IUCN 2003: p. 24).

Teacher Resource Center (TRC), Karachi

TRC has worked for training teachers and have three publications. The centre runs a newsletter with articles for promoting EE in schools.

Bureau of Curriculum and Extension Centre, Quetta

They have initiated a project with WWF to integrate EE into the pre-service teachers' curriculum in 1996. They also developed training materials for PTC and Middle level.

Lead (Leadership for Environment & Development) Pakistan

LEAD is also working towards introducing a school based model covering different aspects of education for sustainable development. LEAD is working on an ESD curriculum.

Child To Child Resource Centre, Directorate of Schools and Literacy, NWFP

The Centre has a regular news letter and focuses on health and sanitation issues. They have also produced teaching manuals.

The World Conservation Union (IUCN)

IUCN has a long association with EE. It has been instrumental in developing the NCS and also Provincial and District Conservation Strategies. It has worked closely with the Education Departments on EE in different provinces. They have a large number of publications which are a rich source of data and issues on the environment in Pakistan. A website www.edu.iucnp.org is also available on Pakistan biodiversity and EE.

PIEDAR (Pakistan Institute of Environment-Development Action Research)

PIEDAR has worked in the field of environment. Their publication includes curriculum and teaching guidelines. They have also conducted in service teacher training, with more than fifty-five schools benefiting.

(continued)

United Nations Children Fund (UNICEF)

UNICEF is working in close collaboration with the Education Department in different provinces and also addresses issues related to environment and health. They have conducted teacher training programmes and produced modules for teachers as well as publishing a newsletter on environmental issues.

A number of organizations are publishing magazines and newsletters for all sections of society and some of the initiatives taken by them are as follows:

- Raising public awareness.
- Teacher training.
- Provision of educational aids.
- Training workshops for conservation in various communities.
- Supplementary materials for curriculum development.

(Shahid et al. 2005: 10)

Environment related education and awareness was included in Pakistan's National Conservation Strategy (NCS). This particular strategy looked into developing awareness and an action plan in the light of an increasingly threatened natural resource base. The NCS has been instrumental in bringing together a number of stakeholders and has facilitated effective coordination between government action, the work of NGOs, the private sector, local bodies and local communities in pursuit of sustainable development. Nonetheless, much remains to be done in the implementation of the NCS action plan.

Development of Mass Awareness Through the NCS

Under the NCS initiative, five key objectives were defined including help to civil society, and the development of environmental awareness; the promotion of basic environmental knowledge and role of people; how to anticipate, avoid and solve environmental problems; how to develop the capacity to contribute and evaluate environmental policies, measures and programs, the development of a sense of urgency leading to direct participation in appropriate action. The focus was mainly on communication strategy and EE where it followed approach of "systematic support for effecting behavioural change" rather than reactive publicized solutions to environmental problems. The authors of the situation analysis study (2005) note that despite several efforts no specific strategy for EE has arisen. The study comments that EE is not considered as part of the continuum of mass awareness. As part of NCS

(continued)

initiative for mass awareness more focus was given to communicating via the mass media than through formal education (IUCN 2000) (http://www.environmental-mainstreaming.org/nssd/country/pakistan/pamtr04.htm. Accessed on November 21, 2014). In the areas of curriculum development, the approach of NCS was to 'revise, restructure and update' with a focus on existing curricula. Pilot materials development projects were developed within the Ministry of Education and EE was included in the Postgraduate Teaching Certificate and Certificate of Teaching curricula. The report visualises that 'General Science' in future will be termed as Environmental Education" (Shahid et al. 2005: 7–8).

The country has launched a Coordinated Environmental Education Project (CEEP) for training teachers, governmental officials and decision-makers. Special programs are run to 'green' opinion leaders and religious leaders. The DEEP (Daudpota Environmental Education Program) Prize for outstanding work in EE has been established. A 16-week training course spread over a period of 2 years is run for government, NGOs, business and military personnel under the LEAD program. Environmental (nature) clubs have been established in schools nationwide (http://pub.iges.or.jp/modules/envirolib/upload/1535/attach/ee-e.pdf. Accessed on November 21, 2014).

The Coordinated Environmental Education Programme (CEEP) is a comprehensive Government initiative to develop a concern for the environment through formal education and was initiated in 1986 by the Ministry of Education and the Environment and Urban Affairs Division, with the assistance of South Asian Cooperative Environmental Education Programme (SACEP) and UNESCO. It was aimed at training teachers, educators, decision makers and planners. It also worked on curriculum research, the production and testing of Environmental Education kits and materials, and better liaison with national, regional and international agencies in this field.

Under the Coordinated Environmental Education Programme (CEEP) in 1986, an effort was made to revise science textbooks. However, this first exercise left much to be desired, and periodical revisions have been needed in textbooks to refine and make the areas of environmental relevance more effective. In most cases private schools, which are now using textbooks prepared abroad, are more willing to add new and contemporary issues like environment to their curricula. It would seem that Environmental Education initiatives as a follow up to CEEP would do well to work with private schools in piloting new environmental curricula and then replicate on a larger scale in government institutions.

The National Education Policy (NEP) 1998–2010 which is the main Policy document for the Formal Education system in Pakistan also serves as a current policy document for greening efforts by the government. The policy statement, according to the NEP, is that for the primary education where enrolment, retention and capacity building primary school teachers was focussed.

"The Ministry of Environment developed the strategy for Mass Environmental awareness and Education in 1996. The objectives of the strategy are described in the seven-step sequence provided below:

- Identify and prioritize environmental issues for the country.
- Identify who can provide the technical expertise.
- Identify the target audience for each priority area.
- Identify environmental messages.
- Identify the most effective and appropriate means of communication.
- Identify organizations, agencies and individuals to carry out the activity and;

Facilitate feedback and monitoring" (Shahid et al. 2005: 8–9).

Central government sets policy but it is for area authorities to interpret and implement. An example is seen in the Environmental Education initiatives in Northern Areas of Pakistan. The Environmental Education background paper of Northern Areas Conservation Strategy (NACS) contains recommendations for capacity building and the facilitation of the mainstreaming of environmental concerns into the education process. The Education Department of the Northern Areas has successfully incorporated environmental concerns in Primary curriculum of class 1–5 under Northern Areas Education Project funded by the United Kingdom Department for International Development (DFID) through the British Council. The EE background paper, recommends the integration of environmental aspects into the curriculum/programmes within the formal and non formal education sectors, the development of resource persons, capacity building of stakeholders and the development of material to supplement the EE activities (IUCN 2003: 13).

10.5 Religious Education, Madarassas and Environmental Education

In Pakistan, the government has a pilot project to persuade religious and opinion leaders to develop environment friendly attitudes and behavior and then include environmental themes in their religious teaching and sermons (IGES 2001: 10). Prior to the colonization of India by the British, madrassas were the only institution, imparting both religious as well the worldly education, especially, to the Indian Muslims. After the mutiny of 1857, madrassas in their vision discarded worldly education as British invention (Key Informant Interview with Wakil Ahmad Khan, former Chairman Madrassa Reform Board). The madrassa tradition continued, after the partition of India in 1947, in both India and Pakistan. Today, madrassas' are confined to religious education. They have flourished in Pakistan being defined in the ordinance for registration of madrassas, issued on December 1, 2005, institutions with boarding facilities concentrating on religious education. Broadly, there are five groups of madrassas in Pakistan, imparting religious education to about three million students. The biggest group belongs to the Daeobandi maslik (sub-identity of sect). Numerically, the second biggest group belong to Barelvi maslik, the third is owned and run by Ahal-e-Tashhi (Shia) and the fourth on the ladder by Ahalay Hadid. The fifth and smallest is owned by Jamiat-e-Islami, a religious political party and is hailed as maslik neural (sect neutral) group of madrassas. By and large, Sunni (one of the two main branches of Islam) students get admission in these madrassas. All five are well organized, having regular and well defined syllabi, curricula, examination boards and certificate or degree awarding criteria and systems.

The 2005 ordinance acknowledged that efforts should be made to introduce reforms in madrassas so as to bring them into the national system. The ordinance provides a framework, according to which a madarassa's Board would comprise educational and religious scholars headed by an official of federal secretary rank appointed by the Prime Minister of Pakistan. However, the madrassas rejected the ordinance soon after its promulgation and have expressed reservations on coming under the government control. Consequently, the madrassa reform board is on paper only and the capacity for such schools to support environmental education is limited.

At the moment there is a vacuum between demand for education and its provision. The madrassas fill the gap. Increasing poverty is a major factor responsible for keeping these institutions in high demand. Poor parents want to put their children in any institution which can provide food, shelter and clothing, and provide education. In the aftermath of 9/11 attacks in the USA, and the invasion of Afghanistan, the international community has showed its willingness to help mainstream madrassas, which it feared could support and proliferate militancy. As a result the International Centre for Religious Diplomacy (ICDR), the British Government and the Turkish government has offered support. Central leadership from all the madrassa groups have paid study visits to UK, Turkey and the US to examine alternative educational systems. This resulted in signing of a Memorandum of Understanding (MoU) on the issue of the nature of education between these groups and the government of Pakistan. The MOU named as 'Agreement between the Government of Pakistan and Ittehad-e-Tanzeemate-Madaris Pakistan (ITPM)' was signed on October 7th, 2010. One of the main clauses of the agreement said; "no Madrassa shall teach or publish any literature which promotes militancy or spread sectarianism, nothing contained herein shall bar the comparative study of various religions or schools of thought or that study of any other subject covered by the Holy Quran, Sunna or Islamic Jurisprudence" (Government of Pakistan and ITPM 2010).

The hope is that the power of the religious schools can be harnessed in the drive towards environmental education and education for sustainable development. As per the NASSD Report of IUCN,

> The representation of religious schools in the interest groups is very limited and will not be enough to provide an opportunity to these institutions to play their role as intermediaries in fostering awareness about sustainable development principles. There is a dire need to initiate linkages with religious institutions and the clergy in order to integrate environmental concerns into their programmes. This can best be achieved by giving them the understanding that environmental messages are not in contradiction with popular Islamic teaching but in fact are reflections of these (IUCN 2003: 18).

10.6 ESD Under the UN System

On March 1, 1992, the Government of Pakistan approved Agenda 21 which is the National Conservation Strategy (NCS), to provide umbrella presence for all other sustainable development initiatives (http://www.un.org/esa/earthsummit/pakis-cp.htm#chap2. Accessed on November 21, 2014). At the federal level it is coordinated informally and involves other agencies like the Environment and Urban Affairs Division, the Sustainable Development Policy Institute (SDPI) and the World Conservation Union, IUCN. For regulation these institutions are linked to the Pakistan Environmental Protection Council. The Pakistan Environment Protection Agency (PEPA) is now a quasi-judicial body. SDPI is instrumental in providing research base for better implementation, helping with technical and policy analysis support. IUCN is instrumental in formulating and adopting the NCS.

Pakistan has made some headway in implementing Agenda 21 by promoting education for sustainable development in particular fields. Establishing a voluntary environment core to monitor afforestation and pollution is an important step. More than 3000 environment clubs have been setup countrywide to provide support to the Mass Afforestation Programme, by educating the public, and ensuring presence as pressure groups (http://www.un.org/esa/earthsummit/pakis-cp.htm#chap36. Accessed on November 21, 2014).

Pakistan was a part of the UNESCO-UNEP International Environmental Education Programme (IEEP). Under the programme the Government has taken steps to reorient education towards sustainable development with a compulsory element in the curriculum at secondary and intermediate level. To promote environment as a subject at the school level recommendations have been made for the preparing text books. A resource center and a library are to be built by the Pakistan Environmental Protection Agency. The programme also proposes the involvement of Non-Governmental Organisations (NGOs) in the advocacy campaign while overall implementation of ESD would be in the hands of local government.

NGOs who will work with the local communities are important and a number are spearheading a new emphasis on education for positive change. The strategies designed to heighten awareness and to foster acceptable attitudes and behaviour toward the environment embrace a variety of non-traditional approaches, some of which appear to be having a positive effect. The ultimate goal is environmentally sustainable development of the region (Shahid et al. 2005: 8–9).

In Pakistan, radio is the best medium for environmental communications. It is particularly effective in reaching two large and pivotal audiences; the users of natural resources who have low literacy levels and a high interest in these resources, and women who are both users of natural resources and processors in subsistence agriculture (Rizvi 1999: 48). Increasing public awareness is at the bottom of promoting education for sustainable development, and a mass awareness campaign has been promulgated through television, radio and the press.

Capacity building requires human capital. It is impossible to promote education for sustainability without first training the teachers. The teachers training 'centers of excellence' will facilitate training of teachers, at the tertiary level. The government is looking forward to introduce a similar programme at the lower levels. Pakistan has also availed multilateral financial and technical support for promoting sustainable development. The CIDA funded Pakistan Environment Programme (PEP) initiated in 1996, is one of those. The $ 29.2 million loan agreement with the World Bank is to undertake the Environmental Protection and Resource Conservation Project (EPRCP). The project was aimed at strengthening government at the federal, provincial and the local level. Rural development was one of the major aims of the project promoting environmental sustainability through education is basic to the initiatives (http://www.un.org/esa/earthsummit/pakis-cp.htm#chap2. Accessed on November 21, 2014).

10.7 Conclusion

Over the last 68 years, Pakistan has reiterated a commitment to the promotion of education in official policy statements at the highest level. The first national education conference was held soon after independence in 1947, to which Father of the Nation, Quied–e-Azam Muhammad Ali attached great value. Since then, about two dozen policies and policy initiatives have been launched. But, almost every initiative has failed in achieving the proposed targets, which resulted in policy failure. The single agreed goal of sustainable development has not been realized as yet, and the government at all levels is looking forward to develop further and sustain it thought capital investment in infrastructure and business schemes, in the real sector and agriculture. Lack of commitment at all levels translated into poor compliance.

Financing is one of the major issues in promoting education in Pakistan. For the first few years, after independence the country spent only about one percent of the GDP on education. And this only increased by a small amount in succeeding years despite the realization that human resource development was the precursor for all other development. However, the national education policy, for the first time in 1992, brought the federal budgetary allocation to 2.7 % of the GDP. Education was viewed as important, but equally important was the fact policies were executed poorly and the dream of translating slogan of 'Education for all,' into reality could not be realized. There seems little incentive in promoting education for sustainable development. However, now with the Millennium Development Goals (MDGs) graduating into Sustainable Development Goals (SDGs), an interconnected development framework needs to be developed at the sovereign level that takes education to lead sustainable development.

References

Environment and Development. (1992). http://www.un.org/esa/earthsummit/pakis-cp.htm#chap2. Accessed on 21 Nov 2014.

Government of Pakistan, Planning Commission. (2007). *Vision 2030*. http://www.pc.gov.pk/vision2030/Pak21stcentury/vision%202030-Full.pdf. Accessed 21 Nov 2014.

IGES (Institute for Global Environmental Strategies). (2001). *Report of the first phase strategic research, environment education*. http://pub.iges.or.jp/modules/envirolib/upload/1535/attach/ee-e.pdf. Accessed on 21 Nov 2014.

IUCN. (2000). *National Strategies for Sustainable Development (NSSD)*. Final report, Islamabad: IUCN Pakistan. http://www.environmental-mainstreaming.org/nssd/country/pakistan/pamtr04.htm. Accessed on 21 Nov 2014.

IUCN. (2003). NASSD (Northern Areas Strategy for Sustainable Development). *Background paper on environment education*, IUCN 2003. https://portals.iucn.org/library/efiles/documents/2003-095_5.pdf

Ministry of Education. (2009). *National education policy 2009*. Islamabad, Pakistan: Government of Pakistan.

Ministry of Environment. (2005). *National environment policy*. Islamabad: Government of Pakistan.

Ministry of Environment. (2011). *National climate change policy draft*. http://www.lead.org.pk/cc/attachments/Resource_Center/NAP/pakistan.pdf. Accessed on 21 Nov 2014.

Ministry of Environment. (2014). *National environment policy 2005*. Government of Pakistan. http://environment.gov.pk/NEP/Policy.pdf. Accessed on 21 Nov 2014.

Ministry of Interior. (2010). *Agreement between the Government of Pakistan and Ittehad-e-Tanzeemate-Madaris Pakistan (ITPM) draft MOU*. Islamabad, Pakistan: Government of Pakistan.

National Implementation of Agenda 21: *Review of progress* made since the United Nations conference on Environment and development, 1992. http://www.un.org/esa/earthsummit/pakis-cp.htm#chap36. Accessed on 21 Nov 2014.

Rizvi, A. R. (1999). Environmental Education in Pakistan. In IGES (Ed.), *Environmental education in the Asia and Pacific Region: Status, issues and practices*. Japan: IGES.

Shahid, S., Mueen, J. A., Zahidi, T. S., & Sahibzada, N. A. (2005). *Situational Analysis Study (SAS) of environmental education in Pakistan*. Islamabad: Curriculum Wing, Ministry of Education, Government of Pakistan.

Spreading Environmental education through 'Green School Programme'. (2014). http://wwf.org.pk/blog/2014/10/31/spreading-environmental-education-through-green-school-programme. Accessed on 21 Nov 2014.

UNESCO. (2012). *Education for all global monitoring report*. http://unesdoc.unesco.org/images/0021/002175/217509E.pdf. Accessed on 21 Nov 2014.

Chapter 11
Education for Sustainable Development for a Peaceful and Sustainable Sri Lanka

Upali M. Sedere

11.1 Introduction

Sustainability is the ability of an individual, a family and community, a nation and its institutions to find the path to development while retaining what has been achieved in overcoming various challenges, problems, and threats. This means that development is essential, and that it needs to be built on achievements made through socio-economic, environmental and cultural means, while acknowledging that development will largely rest on the available environmental resources. It should be emphasized that the concept of sustainability is about achieving and maintaining development on some steady path into the future. A person, a family, a community, a nation or an institutional system could further advance or develop only if what is achieved in terms of development is sustained.

We are at a juncture where environmental degradation, economic instability, inequity and social strife are more pronounced, calling for an integrated approach to address these. The Earth Charter sets forth an integrated approach to addressing the interrelated problems confronting the world community, and asserts that this ethical framework involves respect and responsibility for the community of life, ecological integrity, social and economic justice and equity, democracy, alleviation of poverty, non-violence and peace (UNESCO 2007).

At the International Forum on ESD Dialogue held in Japan in 2008, Matsuura stated that, "The ecological imbalances and socio-economic and financial problems that we face today strongly indicate that our interdependence has become destructive and unsustainable. The urgent challenge is, therefore, how we can learn to generate a new constructive interdependence, as encapsulated in the concept of

U.M. Sedere (✉)
International Consultant on Education, Rector and CEO of the Sri Lanka International Buddhist Academy (SIBA CAMPUS), Kundasale, Kandy, Sri Lanka
e-mail: upalisedere@gmail.com; upalisedere@yahoo.com

R.N. Gorana, P.R. Kanaujia (eds.), *Reorienting Educational Efforts for Sustainable Development*, Schooling for Sustainable Development,
DOI 10.1007/978-94-017-7622-6_11

'sustainable development.' It is the task for ESD to contribute to this global learning process" (UNESCO 2009: 7). In this context, it is important for every world citizen to understand the challenges, problems and threats that hamper sustainable development. Sri Lanka experienced a 30-year civil war initiated by the militant Tamil Tiger movement Liberation Tigers of Tamil Eelam (LTTE) from 1983 to 2009. It faced insurgencies by Janatha Vimukthi Peramuna (JVP) (People's Liberation Front), a predominantly Sinhalese extremist group (in 1971 and again in 1987–1989). The civil war and terrorism seriously hampered peaceful living, brought forth destruction, stalled development interventions and damaged the social fabric and peaceful coexistence of ethnic groups. Many social inequities have been created by such human precipitated disasters.

Sri Lanka was also one of the countries devastated by the earthquake and subsequent tsunami on 26 December 2004. Natural disasters create emergencies and pose serious threats to lives, livelihoods, human existence and properties and they bring to the fore long term social, environmental and economic unsustainability. With global warming, the frequency of natural disasters, disease and threats to the environment are on the increase. In natural and manmade disasters, it is typically the poor that are the worst hit for they have the least resources to cope and rebuild (see http://www.globalissues.org/issue/522/natural-disasters. Accessed on April 6, 2011).

Climate change is real (IPCC 2001) and Sri Lanka is already experiencing the impacts of global warming. The country has experienced unprecedented heavy rainfall, triggering severe floods and life threatening landslides. In 2011, Sri Lanka lost over Rs.30 billion (US $27mn) of rice in the unprecedented floods in the Eastern Province in off season rains (UNDP 2011) (http://www.desinventar.lk/des_html/news_event/news_events.html. Accessed on April 6, 2011).

In the rapidly changing times of today, learning has particular importance for human development--particularly in transmitting rapidly expanding knowledge and confronting challenges created by humans and nature. Education in that sense is a means of conflict resolution and disaster, emergency and threat management, which are so critical for sustainable development. This chapter examines how Sri Lanka has addressed these concerns through education as an integrated and comprehensive approach. Sri Lanka is a case worth exploring to understand how education has contributed to addressing developmental issues like poverty and population growth, despite various challenges (ethnic strife, terrorism, natural disasters) that the island nation has faced and continues to face.

11.2 Issues and Concerns

Sri Lanka has made substantial gains in poverty alleviation, controlling population growth, achieving universal education, and ending civil war but the challenges of inequality, social cohesion, peace, environmental degradation, disaster management are still of much concern. It is well known that these concerns are shared nationally and globally and the Rio (1992), Johannesburg (2002) and Rio+20 (2012) World Summits stand testimony to them.

Like any other developing country, Sri Lanka has economic, environmental and socio-cultural issues that are critical and need to be addressed for spearheading development which is sustainable. This tear-drop shaped island nation is endowed with a salubrious climate, valuable biological resources, abundant water resources, fertile soils, and valuable minerals. Sri Lanka is one of the three biodiversity hotspots in the South Asian region (alongside the Eastern Himalayas in Nepal, northeastern India, and Bhutan; the Western and Eastern Ghats of India and Sri Lanka; and the Indo-Burma hotspot in India and Myanmar). Biodiversity hotspots of the region have been divided into three categories. The highlands of southwestern Sri Lanka along with the Western Ghats of southwestern India are in Category I (UNEP 2009: 31). The country has one of the highest numbers of plants and animal species per unit area and it is amongst the most important biodiversity 'hotspots' in the world due to the tremendous pressure exerted on its limited natural resource base. Recently anthropogenic threats to its forest and other natural resources (e.g., coasts, rivers, soils, fauna and flora) have increased exponentially (Alagan and Aladuwaka 2010) (see http://www.gvglobalvision.org/publications/2020%20Vision%20for%20Gender%20SensitiveDevelopment%20and%20Environment.pdf. Accessed on April 6, 2011).

"The ethno-nationalistic conflict that ended recently continues to have a serious impact on the socio-economic situation in the country, and has inflicted considerable damage to natural resources" (http://www.un.org/esa/agenda21/natlinfo/countr/slanka/nsds.pdf. Accessed on April 6, 2011). The environmental challenges in Sri Lanka can be attributed largely to global industrialization. A significant global impact of industrialization is global warming which has brought pronounced climatic changes in Sri Lanka including changing patterns of rainfall and temperatures that are causing new concerns.

11.3 Environmental Issues and Problems

- Deforestation caused by shifting cultivation, timber exploitation, fuel wood collection, encroachment of agriculture on forests;
- Coastal erosion and resulting damage to beaches, coastal conservations and communities;
- Soil erosion and loss of agricultural productivity of the land;
- Pollution of water bodies, coastal ecosystems and beaches caused by the discharge of industrial pollutants and raw sewage;
- (Protected) sea level rise, landslides, floods and other natural disasters, the victims of which are predominantly from the poorest segments of society, who are compelled to take the risk of occupying vulnerable areas;
- Situation of reservoirs built for irrigation and/or hydro power generation purposes;
- Poor natural resource management and land use patterns, resulting in the marginalization of lands, loss of productivity and depletion of non–renewable resources;
- Atmospheric pollution from vehicle emissions and industries;

- Urban expansion without concomitant infrastructural growth, resulting in the proliferation of slums and shanties, and many other urban problems (see Sri Lanka's Environmental Issues and Problems. http://www.environmentaljournalists.org/sri_lanka_green_issues.html. Accessed on April 6, 2011).

Sri Lanka has had an intractable tryst with ethnic conflict (which broke out in early 1980s) that engendered the civil war between the majority Sinhalese population and the minority Sri Lankan Tamil population who wanted a separate and independent ethnic state. The war went on for over 30 years as the parties involved held non-negotiable preferences and ended controversially in 2009 with the conflict still continuing. The government along with Ministry of Education and leading UN Agencies and Development Organizations like German Technical Cooperation (GTZ) (Deutsche Gesellschaft für Technische Zusammenarbeit) has been working on various approaches to bring about peace and social cohesion.

Disasters, whether natural or manmade cripple a country—more so if it is a developing country. The vulnerability of the country is compounded by indiscriminate and short-sighted economic development, environmental degradation and a host of other social and environmental factors.

> Recurrent disasters, mainly landslides, floods and drought, have caused losses to human life, property and the environment and have often eroded development gains. Though considered as 'natural' disasters, it is known that improper land use and other human activities are predisposing factors. The tsunami of December 2004 resulted in the loss of 31,000 lives, displacement of approximately 443,000 people with more than 900 children being orphaned, and severe damage to peoples' livelihoods, coastal ecosystems and infrastructure. As an emerging economy, the challenge for Sri Lanka is to achieve sustainable high economic growth with greater equity, whilst integrating in the process of globalization, achieving permanent peace and rehabilitating and reconstructing the tsunami affected areas (http://www.un.org/esa/agenda21/natlinfo/countr/slanka/nsds.pdf. Accessed on April 6, 2011).

The Sri Lanka Strategy for Sustainable Development (SLSSD) seeks to achieve this vision through eradication of poverty, ensuring the competitiveness of the economy, improving social development, ensuring good governance, and a clean and healthy environment. Sri Lanka's vision for sustainable development is "Achieving sustained economic growth that is socially equitable and ecologically sound, with peace and stability" (NSDS 2007: iv) (see http://www.rrcap.ait.asia/nsds/uploadedfiles/file/sa/sl/srilanka%20nsds%20report/SL-SDS%20report.pdf. Accessed on April 6, 2011).

11.4 Education and Sustainable Development

Sri Lanka, despite the civil war and the tsunami in 2004, has been able to sustain its overall economic progress because of its educated population. It has a high literacy rate of 94.5 %, with 67 % having received secondary education and over 80 % of the population aged 16–35 with/in secondary education (Sedere 2005). The people of Sri Lanka have sustained not only their personal lives and day to day general living in times of conflict and global economic instability, but also maintained the economic development without allowing it to collapse. The criticality of education in contributing to sustainability becomes clear from the achievements made by the country in addressing problems of population growth and poverty alleviation. In 2002, the Johannesburg Declaration and the Implementation Plan placed a strong emphasis on the need to alleviate poverty for progress to be made towards sustainable development.

It was education that ensured the breaking of the vicious circle: no educational opportunity, no individual development, no economic growth, no building of human capital leading to no national development and then back to no development…no education and so on. As impacts of human activities and natural disasters have escalated so, too, have the risks and threats associated with those impacts. The risks and threat are interrelated and their manifestations are determined by social organization, the natural resource base, the political climate, education and health, and economic progress. Education is the most important intervention for an integrated approach to addressing the interrelated problems confronting the world community and a driver which could facilitate sustainability.

Formal education in Sri Lanka is, for all intents and purposes, a state concern. Since the introduction of seminal educational policies in the 1940s, primary, secondary and tertiary education have been provided in state institutions, free of charge to all Sri Lankan citizens. In addition, these policies helped establish an extensive system of schools throughout the country, bringing education within the reach of all citizens. The system is managed by the Ministry of Education and Higher Education, through the Department of Education. The National Institute of Education (NIE) is the central organization for curriculum development.

11.5 Population, Education, and Sustainable Development

The Johannesburg summit WSSD 2002 identified growing population as one of the key areas and challenges for sustainable development. "According to the 2012 census the population of Sri Lanka was 20,359,439, giving a population density of 325/km2. The population had grown by 5,512,689 (37.1 %) since the 1981 census (the last full census), equivalent to an annual growth rate of 1.1 %" (Department of Census and Statistics 2012) (http://www.statistics.gov.lk/PopHouSat/CPH2011/Pages/sm/CPH%202011_R1.pdf. Accessed on April 30, 2012.) as opposed to 2.8 % in 1970 (see http://www.whosrilanka.org/LinkFiles/Health_Information_EPI_Fact_Sheet_-_Sri_Lanka_2008.pdf. Accessed on April 6, 2011).

Sri Lanka was an early starter on family planning. In 1951, family planning was introduced by the Government of Sri Lanka (then Ceylon), through the formation of the Family Planning Association. The government endorsed Sweden to provide external assistance to the Family Planning Association. Until 1972, population education was not addressed through general education. Educational reforms of 1972 introduced population education as a school subject from Grade 6 through Grade

10. Population education advocated the concept of a small family, its advantages in eradication of poverty; achieving better health; providing better education; achieving better socioeconomic status for the individuals, families, and the nation. Many seminars were conducted on this subject. Parents, School Development Committees and voluntary organization worked hand in hand with the schools.

All pre-service and in-service teacher education programmes took up population education as a priority subject in the school curriculum. Sri Lanka successfully controlled its population growth during 1980–2000 due to many interventions, of which education, school curriculum, methodology of instruction, parent education, etc., particularly helped reinforce the health sector interventions. The success has been even better in the last decade (2000–2010). The 2012 Census reports that the average population growth rate is 0.7 % and the life expectancy for women has increased to 79.5 years. The change from 1970 to 2010 indicates the positive change in population control in Sri Lanka. The impact of educational interventions can be clearly seen when population growth rates—the birth rates and death rates—are correlated with the rate of adult literacy and the education level of the population (Sedere 2008) (http://www.aku.edu/ied/conference2008/programprogrammeme3.shtml. Accessed on April 6, 2011). Sri Lanka has been able to achieve and maintain a low population growth rate due to her educational achievements.

11.6 Poverty Eradication and Education

The WSSD (2002) identified poverty eradication as one of the priority areas for sustainable development and indicated that poverty should be halved by 2015. Poverty is a serious barrier to sustainable development. The poorest people have less access to health, education and other services. Problems of hunger, malnutrition and disease afflict the poorest in society and they are also typically marginalized from society (Shah 2011) (see http://www.globalissues.org/issue/522/natural-disasters. Accessed on April 6, 2011). A vicious cycle of cause and effect is established which is hard to break. Poverty, as identified by WSSD, has major implications for education and vice versa. On the one hand, conditions of poverty relegate educational aspirations to a low priority and, on the other, education is the most powerful intervention to eradicate poverty and sustainable development. The Government of Sri Lanka believes that educating the poor is one of the best possible interventions for overcoming poverty. Such policies as the spread of free education, free health care and free rice rations have contributed very much to poverty reduction. Sri Lanka has a long history of poverty eradication interventions and has substantially and satisfactorily reduced poverty over the last three decades to the present level of 9 % of the population below the poverty line.

Sri Lanka's battle against poverty through offering free (1944) and compulsory education (1988) has been a success story. Since independence from British rule in 1948, Sri Lanka has made steady progress in poverty reduction and in bringing down infant, child, and maternal mortality. For instance, there is a decreasing trend in the population growth rate since 1953 when Sri Lanka recorded the highest population growth rate of 2.8 (in its 140 year history of Census). Accordingly, the population growth rate of 1.2 in 2001 has further decreased to 0.7 in the 2011. This and the above mentioned indicators reflect the mitigation of poverty conditions across Sri Lanka (see http://www.statistics.gov.lk/PopHouSat/CPH2011/Pages/sm/CPH%202011_R1.pdf. Accessed on April 30, 2012).

Sri Lanka introduced a Universal Free Education Policy (Free Education Act 1945) in 1945. The education legislation introduced free education in state primary, secondary, and tertiary levels. However education was not made compulsory by this Act. Although there was no compulsory schooling, yet the government appointed School Attendance Officers to ensure that every child attends school. Furthermore, in 1998, for the first time a Compulsory Education Act was passed in the Parliament (Compulsory Education Legislature 1998). By this Law all children in the age group 6–14 should attend school and should be provided with 9 years (from Grade 1 to 9) of schooling. Sri Lanka is one of the developing nations that offers free and compulsory education up to age 14 and education is free up to Grade 13 or on completion of higher secondary education. Compulsory education is offered in the vernacular media of instruction (Sinhala and Tamil) for all since 1961. The Free Education Policy was backed up with significant government expenditure on education (around 4 % of GDP), owing to favourable economic conditions during the 1950s and 1960s (DFID 2006). Sri Lanka was probably the first developing country to invest such resources in education. By 1990, at the time of the declaration of EFA, Sri Lanka recorded a literacy level of 90 %.

The Government of Sri Lanka (GoSL) concentrated on several strategies for educating all children. GoSL has the following provision for all children:

(i) Free education to all school aged children from Grade 1 through Grade 13. Successful students may continue all the way to State Universities. Over the years the policy expanded to all, irrespective of the ethnicity, religious orientations, incomes, and geographical locations.
(ii) Free School textbooks (introduced in 1980) are provided for every student of Grade 1–11 free-of-charge.
(iii) Free school uniforms (commenced in 1993)—annually two sets—are provided for all students in Grades 1–13.
(iv) Mid-day meals for the children in selected schools in the localities where poverty is a concern.
(v) Strategies to expedite education for impoverished/marginalized communities, such as the estate sector and the war torn areas to mitigate inequity.

GoSL poverty alleviation strategies were very much integrated with children's schooling. These have been implemented for a long time, particularly since 1989. In the 1990s the poverty alleviation programme was called the 'Janasaviya' and subsequent governments continued the programme under a change of names to 'Samurdhi'. The Food and Agriculture Organization states that the movement for alleviation of poverty, brought about a major impact on sustainable development (FAO 2003) (see http://www.fao.org/DOCREP/006/Y5030E/y5030e17.htm.

Accessed on April 6, 2011) and GoSL targeted affirmative actions taken under the poverty alleviation campaigns and coupled all efforts with education, health, housing, labour, women's development, agriculture and rural development ministries of the government not only to make the strategy work but also to sustain achievements. The families receiving special assistance under these interventions had to ensure regular school attendance of their school aged children. Today, across the country there are Special Officers in place to address poverty issues and to work with communities and targeted families. These Officers are also responsible for motivating parents to facilitate their children's education, including school attendance. Sri Lanka has managed to enroll all children in Grade 1 and provide health vaccines to all children on time. These strategies work hand in hand with health, education and Samurdhi interventions. Today, including the very poor, everyone believes in education as the main intervention to improve the quality of life, particularly for children to have a better future than what their parents had.

11.7 Peace, Education Initiatives, and Sustainable Development

War is a state of extreme human rights violations, namely the right to life, and of deep country-wide disorganization, that destroys all social and economic progress. Without peace, no model of development can be made possible. Fostering peace is thus also working towards sustainable development. UNICEF (1999) defines peace education as an essential component of quality basic education and defines it as the process of promoting the knowledge, skills, attitudes and values needed to bring about behavioural changes that will enable children, youth and adults to prevent conflict and violence, both overt and structural; to resolve conflict peacefully; and to create the conditions conducive to peace, whether at an intrapersonal, interpersonal, intergroup, national or international level. The UNICEF document further explains that basic education is a process that encompasses the knowledge, skills attitudes and values needed to live peacefully in an interdependent world. In *The Future Global Agenda for Children – Imperatives for the Twenty-First Century* (UNICEF 1999: 11) there is a commitment to "… ensure that education and learning processes help to form both human capital for economic growth and social capital for tolerance, respect for others and the right of each individual to participate with equality within family, community and economic life;…and to challenge the culture of violence that threatens to destroy family and community life in so many countries". The long drawn ethnic conflict between Sinhala and Sri Lankan Tamils, the Tamils and Muslims, and insurgencies led by Janatha Vimukthi Peramuna (JVP) have considerably undermined harmony, trust, and peace which has evidently impacted development of the country.

11.8 Social Cohesion

In recent decades Sri Lanka has initiated many educational interventions in schools to close the gap and bring better social cohesion through education. Social cohesion is a way for peace. Although inequity was evident during the British period, when the medium of instruction was English, in the private or assisted schools all children of all ethnic groups studied under one roof. Mutual and natural friendships, sharing and caring for each other happened between and amongst children of different ethnic groups. This situation changed when the medium of instruction of secondary schools and higher education institutions changed to the vernacular media. Schools were segregated as Sinhala, Tamil and English schools and Muslim schools. This worked against social cohesion. The natural friendships and social cohesion that existed under multiethnic schools under a single roof was lost. There are still a few of such public and private schools in Sri Lanka where children of all ethnic groups study under the same roof. But most schools are ethnically segregated schools. Some of the efforts and interventions of the government to address social cohesion are described here.

11.9 Education for Conflict Resolution

As early as 1992, to help children learn non-violent ways of resolving disputes, the Ministry of Education, with UNICEF's help, launched a programme called 'Education for Conflict Resolution' (ECR). Initially, a core group of resource persons was trained at the National Institute of Education (NIE) in some of the different forms of conflict resolution being used in other countries. They subsequently adapted these and developed their own methods appropriate to Sri Lanka, producing ten different training manuals aimed at principals, teacher trainers, teachers and pupils (Labouisse 1965), (see http://www.unicef.org/sowc96/fsrlanka.htm. Accessed on April 6, 2011). Today, the international community has the Earth Charter as an instrument which contains fundamental principles for building a just, sustainable and peaceful global society, which further states that to preserve humankind in its integrity, unity and diversity must be reconciled and that the recognition of others is the foundation of all relationships and all peace (see http://www.unescobkk.org/education/esd-unit/earth-charter/. Accessed on April 6, 2011).

ECR incorporated the common Buddhist and Hindu idealism as the basis for educating the young. Both religions preach non-violence and meditation for inner peace of mind. These were utilized for conflict resolution education. A group of 500 Master trainers was trained to spread it to all provinces. The training was provided to educators, school principals, teachers and student leaders. ECR is not limited to a few lessons on 'conflict resolution' but it is integrated into the entire curriculum and importantly without confining it just to schools. In 1995, ECR began a media campaign to extend these ideas to parents and to the community as a whole as well.

11.10 Introducing Tamil and Sinhala as Second Languages

Sri Lanka is a multi-lingual and multi-ethnic society and language has always been a barrier to social cohesion. Particularly due to the 30 year war, opportunities for learning each other's language stopped as there was very little social mobility. Although, Sri Lanka has three official languages – Sinhala, Tamil, and English – it is not compulsory for every child to learn both Tamil and Sinhala. The medium of instruction had been the mother tongue and Tamil students study in Tamil and the Sinhala students study in Sinhala. Mother tongue and English are compulsory subjects in the school curriculum and a student has to take these subjects at the General Certificate of Education (GCE) (Ordinary Level) examination. Though the school curriculum has many co-subjects, Sri Lanka has failed to introduce Sinhala and Tamil as a compulsory second language or as a co-subject respectively for the Tamils and Sinhalese students. However, in recent years Tamil and Sinhala have been offered in schools particularly for achieving better social cohesion and peace and these interventions are clearly included and reflected in the National Action Plan on Education for Peace and Sustainable Development (EPSD) launched in January 2012 (see http://www.govlk.com/content/education-peace-and-sustainable-development-national-action-plan-websitelaunched. Accessed on May 14, 2012). In 1998 the Ministry of Education issued a circular for schools to teach Tamil to Sinhala students and Tamil to the Sinhala students as second language in all schools of Sri Lanka. Programmes were developed by the National Institute of Education to introduce the subject from Grade 3 to Grade 9. In 2003, the curriculum for second language teaching was revised and activity-based learning was introduced. However, the teaching of Tamil in Sinhala schools was possible where there was a teacher who could manage to teach and most of the schools did not have a competent teacher to teach Tamil. The situation was even worse in the Tamil schools in the war-torn North and East. In particular, there was resistance to learn Sinhala and in general teacher scarcity was even greater in these areas. In practical terms it was the Muslims who could manage to teach Tamil to Sinhala students and Sinhala to Tamil students as most of the Muslims speak Sinhala and Tamil. Although the practical difficulties were there, yet the programme continued.

Special teacher training programmes for those who were teaching Tamil or Sinhala as a second language were organized. In recent years the Ministry of Education has created the provision for the recruitment of second language Tamil and Sinhala teachers. However, the provision of recruitment largely depends on provincial administrations and the school and it has not worked out well. Since the defeat of the Liberation Tigers of Tamil Eelam (LTTE) in May 2009 there has been a new momentum and interest in the learning of Tamil and Sinhala in both ethnic communities. The Eastern Province has taken steps to provide teachers and teacher training to strengthen Sinhala and Tamil teaching in the schools. The Northern Province has taken similar steps to teach Sinhala in schools. The recently launched Ten Year National Action Plan on EPSD – Education for Peace and Sustainable Development (January 2012) – also clearly recognizes the second language requirement as one of the seven proposed interventions for peace and sustainable development.

11.11 Social Contract

Over the 30 year war, Sri Lanka had a generation of youth who had no friendly transactions and had developed mistrust. Social mobility from the predominantly Tamil areas to predominantly Sinhala or Muslim areas as well as from the Sinhala areas to the Tamil areas did not occur. The lack of sharing became even more difficult due to the inability to use each other's languages. LTTE particularly cultivated adverse communal images for their advantage. The mistrust and fear of each other by the ethnic groups produced a generation of youth who regarded each other as cultural 'strangers'. The three major ethnic groups of Sri Lanka historically lived in harmony and shared so much of the common culture. All of this was lost or eradicated by the 30 years of bitter civil conflicts. The social contract has been a powerful tool that Sri Lanka has introduced through the schools to cultivate and regain ethnic tolerance for one's fellow citizens, while developing opinions and attitudes of the children, families and communities to the ethnic diversity of the country. Social contracts are to enhance better understanding of each other as people, and citizens of the same country, and to be governed by a common set of social norms and expectations. People are more likely to adhere to social contracts under certain conditions – when they do not consider each other as cultural 'strangers'; that is, when they have more understanding of each other as people, as citizens of the same country or as citizens of a 'similar' country where it is believed that the same norms and expectations govern social contracts.

The idea of social contract has developed well and the programme has had many activities to cultivate such contracts. For example, following two mutual school visits, a few of the students have even made personal visits to each other's homes. The school visits programme has taken on a momentum and continues. The programme was implemented in the Northern, Eastern and Southern provinces during the school holidays and the Ministry of Education, the provincial governments as well as well-wishers facilitated and provided the resources. The activity has now expanded to community visits keeping school at the centre. This has been replicated in many places. The community has agreed to work with the war-torn Tamil villages under the community development work. Such initiatives reflect the mutual understanding developed within the communities. Another example is fund raising undertaken by school children of both Sinhala schools as well as Tamil schools for a common community project in a Tamil village.

11.12 Peace Education at the National Colleges of Education

The key role that education could play in social cohesion was recognized by the country and a national policy on social cohesion and peace education (SCPE) was formulated in 2008. “This policy document on social cohesion and peace education has been formed for the purpose of streamlining the system towards peace” (MoE 2008) (see http://www.moe.gov.lk/web/images/stories/publication/peace_policy.pdf. Accessed on April 6, 2011). Sri Lanka has 18 National Colleges of Education (NCOEs) and the Ministry of Education recruits 3,200 trainees annually to become teachers for various school subjects, providing 3 years of pre-service training. After completion of the training the trainee teachers are appointed as teachers. It is extremely important to impart novice teachers with positive attitudes towards ethnic harmony, the value of unity in diversity and issues of equity at school and in the classroom. The National Institute of Education which oversees all National Colleges of Education and with the assistance of the German Technical Cooperation (GTZ) developed different training packages on social cohesion and peace education for trainees as well as for the academic staff of the National Colleges of Education.

11.13 Education, Equity, and Social Cohesion

Though Sri Lanka has achieved a very high degree of literacy with over 85 % receiving secondary education, the quality of classroom instruction in Sri Lankan schools has much to improve to bring equity in to classroom. Considerable inequality in access to education between rural and urban areas exists. The decisive factor is often one’s socio-economic status. Children of all ethnic groups find better schools if their parents are affluent. However in general, ethnic and language affiliation are factors in inequity.

There are substantial education needs not only in the former war regions in the North and East but also in the remote areas of the country. Classroom lessons are not fully attuned to harness the potentials of children and youth. Disparity is evident in the tea estates, better known as estate sector schools, (the estate sector consists of resident populations in plantations who are ethnic Indians) when compared with the other public schools serving the non-estate sector. Formerly most of the estate sector schools were one-teacher schools and today, due to various interventions and investments, those schools are better equipped with teachers and resources than 20 years ago. Equity is a precondition to be fulfilled for a peaceful coexistence of a multiethnic, multicultural and multilingual Sri Lankan society

The Ministry of Education of Sri Lanka has initiated many social cohesion interventions since 1997 to introduce reforms into the educational system. However, there is a lack of competent and experienced actors at all levels to implement the changes. Neither the teaching staff in the schools nor the professional staff of the

Ministry of Education is competent to pave the path to a peaceful future for the schoolchildren and youth

GTZ with financial support of the German Economic Cooperation and Development (BMZ) had been supporting the Ministry of Education in Sri Lanka and the Provincial Councils particularly in the North, East and the Central provinces. In the North and East where the LTTE activities have eroded the social cohesion in Tamil population MOE with GTZ support introduced a wide range of social cohesion activities. Similarly in the Central province where the estate sector schools are many the Tamil population needed additional support to achieve better equity.

The Education for Social Cohesion Programme – (ESC) with GTZ support – developed skills and competencies in children and youth for a peaceful and responsible coexistence. The school environment provides an opportunity for students to learn about tolerance, mutual understanding, human rights, cultural respect, peaceful conflict transformation, and peace. The project operates at three levels of implementation covering a wide spectrum of the education landscape. At national level, it engages with the Ministry of Education, the National Institute of Education and the National Education Commission and advises on policies related to education, further capacity development of management and professional staff, curriculum development and quality assurance. At provincial level, ESC carries out training at National Colleges of Education, Teacher Training Centres and Provincial Departments of Education. At school level ESC engages with 200 schools, community learning centres, teacher centres and communities in an effort to develop innovative concepts and material to meet the needs of teachers, school children and the community.

11.14 Environmental Education and Sustainable Development in Formal School System

Sri Lanka has taken legal steps, policy directions and institutional interventions since 1980 and developed a national environmental plan of action for environmental protection in 1992. National Environmental Policies and Strategies were formulated and a separate Ministry to deal with the environment was established. Collective action has been taken by inter-ministerial committees where the Ministry of Education, the National Institute of Education and the school system are identified and mobilized as key institutions for action (MoE&NR and UNEP, 2009). In 2003, the 'Caring for Environment – Path to Sustainable Development' programme was initiated. It identified forest depletion, haphazard waste disposal, land degradation, uncontrolled sand mining, air pollution, climate change, loss of bio-diversity, pollution of land water, and pollution of marine and costal ecosystems and unsustainable utilization of natural resources as priority areas.

Conservation of the environment is deeply embedded in Sri Lankan society. Ancient Sri Lanka always maintained the ecological checks and balances to face natural threats on sustainability of the lives and livelihoods of the people. In turn people always protected the environment. The curriculum and textbooks as well as the philosophy and the ideology of Sri Lanka's education system have a major objective of conservation of its natural environment. These are well reflected in all educational and curriculum reforms since 1932. In that year, "a comprehensive ordinance covering all aspects of education to lay the foundation for a national system of education" was initiated and was finalized in 1939. To date, it "remains the basic law of Education in Sri Lanka" (MHRDECA 2004: 3). The school curriculum and the teacher education curriculum have environmental studies as a separate subject and it is also integrated with all other school subjects where integration is meaningful and possible. The 1972 primary education reforms were highly environment-oriented and integrated with prevalent public health issues and the quality of life of the rural population. However, with the escalation of ethnic conflict in 1983, issues of social cohesion, environmental destruction, insecurity, economic sustainability became critical.

In 1992, a Report of the National Education Commission (NEC) identified five sets of basic competencies to serve as a means of attaining nine national education goals. These competencies are related to communication, environment, ethics and religion, play and the use of leisure, and learning. They are built into the school curriculum and textbooks prepared by the Ministry of Education, Sri Lanka. In 1992 environmental studies was introduced as a compulsory core subject up to Grade 9 in the school curriculum.

11.15 Textbooks and Environmental Issues

The primary education curriculum from Grade 1 to Grade 5 has 16 themes on environment-related activities which include plants, animals, water, earth, sky, rain, shelter, food, safety and security, beings around us, human behavior, cleanliness, garbage disposal and our heritage. These themes are integrated with science, physical education, aesthetic education and creative activities. The biological environment, man-made (physical) environment and social environment are identified as three main domains to be addressed for sustainable development. The biological environment-related theme 'Plant and Trees' is of particular interest as it directly relates to the immediate environment of the child and is one of the most popular themes amongst the other 15 themes.

Man-made environment is included in Grades 3–5 with a theme on 'Safety and Care' which addresses disaster related concerns. The disasters are also divided into two sub-themes: natural disasters and man-made disasters. The theme is activity-based and children develop knowledge and understanding, attitudes and acquire preparedness-related competencies and skills. For the social environment three themes are included in the primary curriculum: i.e. 'We are Sri Lankans'; 'Our Wellbeing'; and 'Human Behaviour'. The theme on 'We are Sri Lankans' addresses the social cohesion issues and concerns. These are similarly activity-based with coverage of ethnic and religious groups, their cultures being addressed to develop mutual respect among all.

The theme of 'Our Wellbeing' addresses spiritual and social issues and concerns. Human behaviour addresses personal understanding, listening to others; mutual respect, self-respect issues and concerns are covered through various activities. All schools in Sri Lanka are involved and take part in environmental protection programmes. Many schools also have 'Environment Clubs' and work with their communities on environmental issues. The Environment Clubs are registered with the Environmental Authority (EA) of the Ministry of Environment and Natural Resources. The Regional Offices of the EA and schools work hand in hand on environmental concerns and issues. Tree planting campaigns, school landscaping to stop erosion and provide better water management and drainage, develop better sanitation and garbage disposal systems by sorting of types of garbage, developing school policies to stop using playthings in schools, better use of electricity, water and other resources are some of the activities in which these clubs are engaged.

All primary teachers have received training on the 16 themes on sustainable development described under the section above on curriculum. The Ministry of Education's programme is also assisted by the development partners particularly GTZ, UNDP, UNESCO, UNICEF, PLAN and those in the private sector such as MAS Holdings of Sri Lanka. These have made concrete interventions to develop teachers, students and teacher trainers to address environmental issues and develop learning activities and toolkits for sustainable development through education.

11.16 School Curriculum and Textbooks

Sri Lanka introduced school curriculum reforms in 1972, 1980, 1996, 2002 and 2007 with an 8 year revision cycle. In the past, even at primary level, the content of subjects received the greatest emphasis on sustainable development. The new curriculum is focused on competencies that children are expected to have acquired at the end of their general education. The present primary curriculum establishes a comprehensive set of basic competencies in communications, ethics and religion, environment, learning to learn, enjoyment and leisure, preparation for the world of work and personality development. Entry competencies are identified, using specially designed assessment instruments, enabling the early Grade teacher to cater more effectively to children's individual needs. Furthermore, the essential competencies that children are expected to have mastered at the end of each key stage are identified. Teachers are encouraged to ensure that at the end of each unit almost all children in their class have reached the mastery level in the essential competencies, with special emphasis on first language and mathematics. Methodology is more activity-based and assessment is informal and school-based. With all these efforts about 18 % of primary school children do not achieve the expected levels.

As stated in the National Report of Sri Lanka by the Ministry of Education (2004) the Junior Secondary Curriculum from Grade 6 to 9 is a common curriculum comprising nine subjects: first language, English, mathematics, science and technology, social studies, life competencies and citizenship education, religion, aesthetics, health and physical education. The teaching of a second national language, i.e. Sinhala for Tamil speaking students and Tamil for Sinhala speaking students, is introduced at this level where teachers are available. The second national language is taught in two 40-min periods in a week. The environmental studies are given a very heavy emphasis in Grade 6 allocating nine 40-min periods a week. These are affirmative actions taken by the Ministry of Education to ensure that the students of the age group 11–14, as they enter the formal thinking stage of development, acquire the right values, attitudes and thinking for a multiethnic and multireligious society to achieve coexistence, and live in harmony with mutual respect to all. These were affirmative steps to address sustainable development contributing to peace, social cohesion, ethnic and religious harmony and environmental protection.

The Senior Secondary Curriculum from Grades 9 to 11 leading to G.C.E. (Ordinary Level Examinations) and Grades 12 and 13 leading to G.C.E. (Advanced Level Examinations) are more focused on tertiary education requirements. However, particularly at Grades 10 and 11, the G.C.E. Ordinary Level curriculum includes six co-curricular subjects (first language, mathematics, science, religion, history and English) and a student has to select three other optional subjects one from each of the three clusters (baskets) of subjects. These optional subjects include many of the life skills and citizenship-related subjects; health and physical education related subjects and technological subjects. All school textbooks and syllabuses have important competencies and content included on sustainable development. Particularly, essay writing in languages includes such matters as ethnic and religious harmony and social cohesion; disaster management, climate change and environmental degradation with essential competencies to be demonstrated. The content in all subjects where there is a natural link with these concerns are always included. Beside these the co-curricular activities schools concentrate very much on the acquisition of life skills and developing attitudes towards issues relating to environment and human total personality development (Ministry of Education 2004).

11.17 Disaster Education

The Indian Ocean tsunami in 2004 resulting from the earthquake on December 26, 2004 had a devastating impact on the South and Eastern coastal belt in Sri Lanka, killing over 30,000 people. This added a further strain on children and the education sector in Sri Lanka. Thousands of children were killed and the waves leveled or damaged 182 schools, affecting an estimated 100,000 children (see http://www.unicef.org/srilanka/activities_881.htm. Accessed on April 6, 2011). School safety is a new concern in Sri Lanka. With the recent experience of tsunami, landslides and floods steps are being taken to provide school safety to children. Schools in the coastal areas go through tsunami drills with the tsunami alarm systems that are now installed and in operation.

The National Institute of Education and UNDP worked together in year 2010 in developing a series of five Booklets as supplementary readers – Landslides, Floods, Cyclones, Earthquakes and Tsunami – on 'Disaster Management' for dissemination in schools. Teachers were trained by NIE in their effective use. The Provincial Education Ministries have also developed specific strategies in addressing disasters that are more common in their provinces. GTZ, is another organization that works with NIE and the Provincial Administrations to prepare the schools' teaching staff and the staff of the Ministry of Education for disaster preparedness and management as well as psycho-social counseling for children and youth. Besides these initiatives, disaster preparedness and management have also been integrated into the school curriculum. Children and teachers along with DMC were involved with the identification of the safe routes and safe evacuation sites through 'transect work' and developing 'Hazard & Evacuation Maps'. Before the mock drills, a separate awareness programme was conducted to educate teachers and students on the measures to take place.

The UNDP project contributed in terms of intensive policy advice; preparation of professionals and high government officials in disaster management; elaboration of curricula for children with disabilities and youth; and psychosocial counseling to overcome traumatic experiences and improve social integration. GTZ, MOE, NIE and Asian Disaster Preparedness Center (ADPC) are working together on an Education for Social Cohesion, Disaster Risk Management and Psycho–Social Care Programme. These programmes have been well received by schools and communities and have made noteworthy progress towards sustainable outcomes in disaster management.

11.18 Conclusion

To a great extent, the general progression of education of Sri Lanka can be seen as taking an education for sustainable development approach. Everyone, including the poor, believes that education is the means to improve the quality of life, particularly for children to have a better future than what the parents had. Seminal legislation like free and compulsory education, supporting initiatives like bearing of direct and indirect costs of schooling by the government, and monitoring of education have actually improved education delivery. Education policies and systems are geared to address environmental, social, and economic problems as a means to address development challenges. Education is seen as the most powerful intervention to eradicate poverty and achieve sustainable development as it is well recognized that poverty relegates educational aspirations to a low priority. Another major aspect that education can achieve is reconstruction after the bitter ethnic conflict which has ended recently. Achieving harmony and peace is one of the objectives of the National Educational Action Plan on Education for Peace and Sustainable Development launched in the year 2012. To be specific, the content, method and strategies of education are tuned to the implication and consequences of ethnic conflict, language issues, population control, poverty alleviation, social cohesion, disaster risk reduction and management. Education will be a chief vehicle for achieving Sri Lanka's vision for sustainable development, "Achieving sustained economic growth that is socially equitable and ecologically sound, with peace and stability" (see http://www.un.org/esa/agenda21/natlinfo/countr/slanka/nsds.pdf. Accessed on April 6, 2011).

References

Alagan, R., & Aladuwaka, S. (2010). *2020 vision for gender sensitive development and environment.* http://www.gvglobalvision.org/publications/2020%20Vision%20for%20Gender%20SensitiveDevelopment%20and%20Environment.pdf. Accessed on 6 Apr 2011.

Department of Census and Statistics. (2012). *Sri Lanka department of census and statistics (2012). Census of population and housing 2011, enumeration stage February – March 2012: Preliminary report 2012.* http://www.statistics.gov.lk/PopHouSat/CPH2011/Pages/sm/CPH%202011_R1.pdf. Accessed on 30 Apr 2012.

DFID. (2006). *Educational opportunities for the poor: Sri Lanka.* www.odi.org/resources/docs/4074.pdf

FAO. (2003). *Sri Lanka case study Samurdhi programme.* http://www.fao.org/DOCREP/006/Y5030E/y5030e17.htm. Accessed on 6 Apr 2011.

http://www.un.org/esa/agenda21/natlinfo/countr/slanka/nsds.pdf. Accessed on 6 Apr 2011.

IPCC. (2001). *Climate change 2001: Synthesis report. A contribution of working groups I, II, and III to the third assessment report of the integovernmental panel on climate change* [Watson, R.T. and the Core Writing Team (Eds.)]. Cambridge University Press. http://www.grida.no/publications/other/ipcc_tar/. Accessed on 6 Apr 2011.

Labouisse, H. (1965). *How Sri Lanka educates children for peace.* http://www.unicef.org/sowc96/fsrlanka.htm. Accessed on 6 Apr 2011.

McConnell, J. (1979). *History of the original earth charter – 1979* http://www.earthsite.org/EarthCharter.htm. Accessed on 6 Apr 2011.

Ministry of Education. (2004). *National report: Development of education in Sri Lanka.* Battaramulla: Ministry of Education.

Ministry of Education. (2008). *National policy and a comprehensive framework of actions on education for social cohesion and peace (ESCP).* Social Cohesion and Peace Education Unit, Ministry of Education. http://www.moe.gov.lk/web/images/stories/publication/peace_policy.pdf. Accessed on 6 Apr 2011.

Ministry of Education & SLNCU (Sri Lanka National Commission for UNESCO). (2012). *Ten year national action plans on EPSD – Education for peace and sustainable development.* (NPESD) UNESCO. (Unpublished) http://www.govlk.com/content/education-peace-and-sustainable-development-national-action-plan-websitelaunched. Accessed on 14 May 2013.

Ministry of Human Resource Development, Education & Cultural Affairs (MHRDECA). (2004). *Education for all, national action plan Sri Lanka.* http://planipolis.iiep.unesco.org/upload/Sri%20Lanka/Sri%20Lanka%20EFA%20NAP.pdf. Accessed on 6 Apr 2011.

MoE&NR (Ministry of Environment and Natural Resources) & UNEP (United Nations Environmental Programme). (2009). *Sri Lanka environmental outlook.* Bangkok: UNEP-RRCAP.

NSDS. (2007). *Mapping of sustainable development strategies in countries emerging from conflict: Asia and the Pacifics.* https://sustainabledevelopment.un.org/content/…/1498mappingasia.pdf

Sedere, M. U. (2005). *Context of educational reform then and now: 21st annual C. W. W. Kannangara memorial lecture*. Colombo: Ministry of Education.
Sedere, M. U. (2008). *Impact of educational reforms on socio-economic development*. http://www.aku.edu/ied/conference2008/programprogrammeme3.shtml. Accessed on 6 Apr 2011.
Shah, A. (2011). *Natural disasters*. Global issues. http://www.globalissues.org/issue/522/natural-disasters. Accessed on 6 Apr 2011.
Sri Lanka Strategy for Sustainable Development. Ministry of Environment and Natural Resources. February 2007 http://www.rrcap.ait.asia/nsds/uploadedfiles/file/sa/sl/srilanka%20nsds%20report/SL-SDS%20report.pdf. Accessed on 6 Apr 2011.
Sri Lanka's Environmental Issues and Problems. http://www.environmentaljournalists.org/sri_lanka_green_issues.html. Accessed on 6 Apr 2011.
UNDP. (2011). *Disaster information management system in Sri Lanka*. http://www.desinventar.lk/des_html/news_event/news_events.html. Accessed on 6 Apr 2011.
UNESCO. (2007). *Highlights on DESD progress*. 2007. http://www.desd.org/Highlights%20on%20ESD%20progress-%20April%202007.pdf. Accessed on 6 Apr 2011.
UNESCO. (2009). *ESD currents: Changing perspectives from the Asia-Pacific*. Bangkok: UNESCO Bangkok.
UNEP. (2009). *South Asia environmental outlook*. ISBN: 978-92-807-2954-2.
UNICEF. (1999). *The future global agenda for children: Imperatives for the twenty-first century, united nations, economic and social council*. http://books.google.co.in/books/about/The_Future_Global_Agenda_for_Children.html?id=kGm1YgEACAAJ&redir_esc=y. Accessed on 6 Apr 2011.
UNICEF. http://www.unicef.org/srilanka/activities_881.htm. Accessed on 6 Apr 2011.
United Nations Environment Programme and Development Alternatives. (2008). *South Asia environment outlook 2009: UNEP, SAARC and DA*. Nairobi: UNEP.
World Health Organization, South East-Asia Regional Office. (2009). *Sri Lanka 2008 EPI fact sheet*. http://www.whosrilanka.org/LinkFiles/Health_Information_EPI_Fact_Sheet_-_Sri_Lanka_2008.pdf. Accessed on 6 Apr 2011.
WSSD. (2002). *Global challenge, global opportunity: Trends in sustainable development trends in sustainable development. United nations department of economic and social affairs*. http://www.un.org/jsummit/html/documents/summit_docs/criticaltrends_1408.pdf. Accessed on 6 Apr 2011.

Chapter 12
Lessons Learned from South Asia

Rajeswari Namagiri Gorana and Preeti Rawat Kanaujia

Having reached the last chapter of this book on Experiences from South Asia, it is necessary to reiterate that the South Asia region is an expansive and populous one with great diversity. Not only does it display a wide variety of races, languages and religions, it is also characterized by great differences in community and personal wealth, economic opportunities, and access to those tangible and intangible phenomena for which many people aspire.

As the chapters have elucidated diversity, not simply biodiversity, is a recurring feature in culture, language, religion (including EE and ESD approaches) which is a strength and equally a challenge. The authors have established the contexts for education to gain a sense of how functional economies are built in the region while overcoming issues of food shortages, poverty, equity, population growth, literacy, cohesion, peace and development seems like an ongoing process. The circumstances for nation building after long periods of colonization have been challenging for the colonized countries of the region.

R.N. Gorana (✉)
Programme Coordinator, Children's Media Unit, Centre for Environment Education (CEE), Ahmedabad, India
e-mail: rajeswari.namagiri@ceeindia.org

P.R. Kanaujia
Senior Programme Coordinator, Northern Regional Cell, Centre for Environment Education (CEE), Lucknow, India
e-mail: preeti.rawat@ceeindia.org

R.N. Gorana, P.R. Kanaujia (eds.), *Reorienting Educational Efforts for Sustainable Development*, Schooling for Sustainable Development,
DOI 10.1007/978-94-017-7622-6_12

12.1 South Asia's Perspective

Needless to say, diverse human features, expressed in a multitude of cultural, social and economic phenomena have consequences for education, including for EE and ESD. It has also been established that there are a number of methodologies and ideologies relevant for schooling for sustainability. We have seen that environmental, developmental and educational contexts form the basis of EE and ESD interventions and practices, many of which appear diffused (as mentioned elsewhere in this chapter, the nature of practice needs to be brought within a deeper understanding framework) and highly contextual ranging from formal, non-formal to informal – not necessarily pointing to any norm at a country level and least at the regional level. The experiences from the region clearly indicate that EE and ESD should not be measured against a global norm. To add, any discussion about education in the nations of South Asia, as touched upon in the previous chapters, is set in the specific physical and human contexts which were well captured, for example, in the Millennium Development Goals promulgated in 2000:

- poverty and hunger;
- universal primary education;
- gender equality and the empowerment of women;
- child mortality;
- maternal health;
- diseases;
- environmental sustainability;
- a global partnership for development.

In South Asia, education has to have a particular resonance with development challenges, making it eclectic. As an extension, the eclectic nature of EE and ESD content should be seen not as a weakness. It only raises the rhetorical question whether efforts that emphasize different aspects and issues of environment and development should be categorized as EE or ESD.

Conventionally, education is associated with schooling in South Asia, but formal schooling is not necessarily the principal educational provision. The informal and non-formal sectors are equally significant if not more significant for many social, religious and artisanal communities; rural people; economically and socially underprivileged, for the illiterate. The challenge and opportunity for EE and ESD professionals and the governments continues to be to work with all the sectors, harness the potential of these sectors so that education for all, in a lifelong frame of reference can be achieved.

As for formal education, access and quality will continue to be overriding concerns. Providing schools with qualified and effective principals and teachers with adequate teaching resources is one of the many challenges to begin with. Where schools have been provided, the drive has been, in the first instance, to enroll children, and especially girls, and then to encourage them to attend classes. In terms of curriculum priorities, literacy—reading, writing and arithmetic—are the least if not the last on the list. Any EE and ESD initiative has to take cognizance of these, and where possible clearly aid in realizing them to achieve the goals of EE and ESD.

In 2014 the UNESCO Decade of Education for Sustainable Development ended and the Final Report was built around ten trends that had been identified over the decade (Buckler and Creech 2014). Clearly, from a South Asian perspective, these need to be read against the contemporary environmental and cultural contexts and the persistent historical characteristics that define the region.

The authors of the Report review their findings under a series of headings and sub-headings:

ESD, an enabler for sustainable development

1. Education systems are addressing sustainability issues
2. Sustainability development agenda and education agendas are converging

Importance of stakeholder engagement for ESD

3. Political leadership has proven instrumental
4. Multi-stakeholder partnerships are particularly effective
5. Local commitments are growing

ESD is galvanizing pedagogical innovation

6. Whole-institution approaches practise ESD
7. ESD facilitates interactive, learner-driven pedagogies

ESD has spread across all levels and areas of education

8. ESD is being integrated into formal education
9. Non-formal and informal ESD is increasing
10. Technical and vocational education and training advances sustainable Development (p. 9)

EE and ESD has to become part of education (as generally defined and understood) at a meta level such that the assimilation itself states the unstated aims of education for a "different planet, arguably a different human nature, and a global culture that is evolving faster than people can comprehend and adapt to" (State of the World 2010: p. 76).

12.2 Recognizing the Challenges for Schooling for Sustainability

The challenges, of the region, are as much for education as for sustainability. Those denied an education suffer enormous social and economic disadvantages. They are amongst those with the poorest health, lowest housing standards, and poorest employment prospects in the world. In fact, they have less of nearly everything in life, except children (UNESCO 2002: p. 13). Governments in the region have shown strong commitments, on paper if not in practice, to achieve Education for All. Education must be seen as the primary agent of transformation not just for economic productivity alone but towards sustainable development, increasing people's capacities to transform their visions for society into reality. In this context, achieving the targets of Education for All remains the biggest challenge for South Asian countries.

As an imperative, the essential goals of education have to be met – learning to know, to do, to be (i.e., to assume one's duties and responsibilities) and to live together with others, as outlined in *Education: the Treasure Within*, the report of the Independent Commission on Education for the twenty-first century Report published in 1996 by UNESCO. It is, thus, not only the foundation for lifelong learning, but also the foundation for sustainable development. Hearn, in 'Amartya Sen's Development as Freedom: Ten Years Later', Policy & Practice: A Development Education Review, refers to Sen's analysis that equates capability with choice. "Education, for example, is crucial beyond its role in production; its most important role being that of increasing human capability and therefore choice." He mentions that Sen cites Adam Smith "who links productive abilities to lifestyles to education and training, and presumes the improvability of each" (2009, p 9–15).

Unfortunately, there are many barriers ranging from such issues as large class sizes, lack of resources for teachers and students, top-down, textbook-centered teaching strategies, in achieving education for all in letter and spirit. Further, all learning is dominated by competitive academic curricula which prioritizes end-of-course examinations which discourage the development of locally and personally relevant learning.

Under these circumstances, it is appropriate to ask, if EE and ESD stand a chance? Experience, interpretation, dispensation in the countries of the region shows a recognition and obligation towards EE and ESD. In fact, without effecting fundamental changes in educational policy, curriculum, teacher training, infrastructure, EE and ESD has found practitioners.

A ray of hope for the South Asia region is the broad and contextual interpretation of ESD which for instance has been clearly articulated in the case of Sri Lanka. In countries like India which have cleared the policy ground through the Supreme Court ruling making EE mandatory at all levels of formal education there is the potential for introducing ESD.

Importantly, ESD in the region needs to be seen as a co-benefit because of the multiple objectives of education (literacy, quality education, EE and ESD), learning approaches (formal, informal, non-formal), areas of learning (peace, cohesion, climate change, happiness, resource conservation and management).

Non-formal education (NFE) (including raising public awareness) has been recognized as a key instrument way back in 1992. There are enough experiences from the region in NFE. "Formal educational systems are notoriously slow to change. Non-formal channels are capable of delivering new information and tapping new approaches and methods for teaching and learning more easily. In professional educational circles, it is the promise of the non-formal sector which is being seen increasingly as a complement to formal education for the future, making the role of non-governmental organizations and associations of all kinds, as well as the media and the arts, ever more important (ECOSOC 2001: p. 7). Non-formal education, including public awareness, may deliver results in a shorter time-frame but would be essential in achieving sustainable future (p. 8)".

Initiatives like the National Green Corps (NGC) in India and Green Clubs in the Maldives have successfully utilized non-formal spaces in formal education through an Eco Club approach. These have opened up the possibility of focusing on the process of learning rather than the content. They have fostered cross curricular learning and the acquisition of twenty-first century skills. Through the non-formal route, EE and ESD have found a place in the schools if not in the classrooms (through the formal curriculum). However, this dispensation runs the risk of being considered suboptimal as it is not a curricular priority at the school level thereby never raising the profile of EE or ESD. As pointed out in the International Conference on Environmental Education in the Asia-Pacific Region, the absence of external examinations for EE or ESD has contributed to lack of status in schools. "Therefore, it is not surprising to find that many teachers, students and parents do not perceive it as a curriculum priority, or that it is difficult to attract teachers to become involved" (IGES and Environment Agency 1999: 66). It is important that such initiatives are developed and anchored appropriately in the curricular system.

12.3 Strategies and Considerations

EE and ESD should emanate from the social, cultural and economic realities of the people. It should not be an endeavour to aspire and strive to follow the development model of the west or result in an obsession with materialism. In fact, EE and ESD should wean away the countries of the region from development-by-imitation. As Sarabhai points out, "If India and other developing countries follow the example of the countries that are considered developed, adopt similar lifestyles, produce and consume in similar ways, it is estimated that by 2050, about four planets would be required to meet human requirements!" The development choices, he says "need to be bold, we need to 'pick right' and avoid temptation of imitation. We need to lead, rather than follow, as we go in the direction of sustainable development. In this we need to re-look at, and often re-learn from, our own past, both in terms of values and practices" (Sarabhai 2010: p 69).

The pedagogy of EE and ESD should be based on oral cultures, made available in mother tongue or local languages and should move beyond formal structures. It needs to be acknowledged that knowledge is not a unitary concept. There are different kinds of knowledge as well as different ways of knowing and learning. The Chapter on Sri Lanka by Sedere in this book clearly indicates that if formal schooling is offered in the mother tongue it is helpful not only in performance but in achieving better social cohesion and peace.

The region is replete with shining examples, at the various levels, that have been launched proactively or in response to a particular environmental or developmental issue. For instance, there is a proactive community and civil society in Bangladesh and India and an activist judiciary, a progressive enviro-legal jurisprudence and an unparallel culture of public interest litigation in India. The focus of several educational initiatives, in Sri Lanka, has been on peace and social cohesion. The learnings can be channelized to capture the imagination of the agency of governments and community based organizations for EE and ESD. The legitimacy of these initiatives needs to be acknowledged and streamlined under ESD.

The ground gained in countries like India, Bangladesh, Sri Lanka, Maldives and Nepal has to lead to a "coherent plan for progression in environmental education or education for sustainability from kindergarten to college level in the formal education sector" (IGES and Environment Agency 1999: 66).

Teacher preparation is critical to creating a cadre of teachers committed to EE and ESD. It is a building block for strengthening the processes of learning such that teachers can support the widening mandate of education to include EE and ESD. In an early report on this theme, Wilke et al. (1987) stated that: The key to successful environmental education is the classroom teacher. If teachers do not have the knowledge, skills and commitment to environmentalise their curriculum, it is unlikely that environmentally literate students will be produced (p. 1). This requires top down changes from policy makers, school administrators and teacher education institutions.

In most countries in the region, EE and ESD in formal education remains embedded within a pre-Rio conception which suggests that EE is best taught through science and/or that conservation approaches tantamount to education for sustainability. EE should not limit itself to such dispensation and include social, cultural and economic aspects and importantly bring about behaviour change. It should foster action and not simply sporadic changes in individuals but a cohesive and collective change.

Government wide commitment for EE and ESD should be garnered from various departments and ministries through the common but differentiated responsibility approach. The principle of common, but differentiated, responsibilities needs to be made into a strategy of ESD. ESD neither vests nor can be vested solely and wholly to formal educations systems or to ministries looking after environment, forests. Some options for a nuanced model of differentiation in terms of both responsibility and capability across government wide ministries/departments needs to be acknowledged and worked out.

To conclude, the growing body of formal education work in South Asia related to EE and ESD has covered a wide range: from awareness to taking action (building capacities), co-curricular to curricular processes. It is very important to derive the patterns from this body of work to consolidate EE and ESD practices and work for a progression. In addition, the diffused nature of practice itself needs to be brought within a deeper understanding framework to aid system-level shifts so that this gets resonated in policies and practices related to formal education, development initiatives, environment and economic matters. Such a progression will, hopefully, contribute to giving a cohesive shape to the deliberative but pragmatic nature of EE and ESD work in the South Asia region.

References

Buckler, C., & Creech, H. (2014). *Shaping the future we want: United nations decade of education for sustainable development (2005–2014)* (Final report). Paris: UNESCO.

ECOSOC. (2001). *Education and public awareness for sustainable development*. Report of the Secretary-General, UN Economic and Social Council. http://www.iatp.org/files/Education_and_Public_Awareness_for_Sustainable.pdf. Accessed 21 Apr 2016.

International Conference on Environmental Education in the Asia-Pacific Region 27–28 Feb, 1999 Yokohama Organized by Institute for Global Environmental Strategies And Environment Agency, Government of Japan.

O'Hearn, D. (2009). Amartya Sen's development as freedom: Ten years later. *Policy & Practice: A Development Education Review*, *8*, Springer. http://www.developmenteducationreview.com/issue8-focus1. Accessed 9 Dec 2015.

Sarabhai, K. (2010). ESD in a developing nation. *Tomorrow Today*. Paris: UNESCO (Published by Tudor Rose on behalf of UNESCO).

State of the World. (2010). *What is higher education for now?* By David Orr. http://blogs.worldwatch.org/transformingcultures/wp-content/uploads/2010/12/What-is-Higher-Education-for--Now-Orr.pdf. Accessed 9 Dec 2015.

UNESCO. (2002). *Education for sustainability – From Rio to Johannesburg*: *Lessons learnt from a decade of commitment*. A Report on the lessons learnt about the contribution of education to sustainable development over the decade between the United Nations Conference on Environment and Development in 1992 and the World Summit on Sustainable Development (WSSD).

Wilke, R. J., Peyton, R. B., & Hungerford, H. R. (1987). *Strategies for the training of teachers in environmental education*. UNESCO-UNEP International Environmental Education Programme, Environmental Education Series No. 25, Paris: UNESCO Division of Science, Technical and Environmental Education.

MIX
Papier aus verantwortungsvollen Quellen
Paper from responsible sources
FSC® C105338

If you have any concerns about our products,
you can contact us on
ProductSafety@springernature.com

In case Publisher is established outside the EU,
the EU authorized representative is:
Springer Nature Customer Service Center GmbH
Europaplatz 3, 69115 Heidelberg, Germany

Printed by Libri Plureos GmbH
in Hamburg, Germany